ON THE BACKS OF OTHERS

ON THE BACKS OF OTHERS

Rethinking the History of British Geographical Exploration

EDWARD ARMSTON-SHERET

University of Nebraska Press
Lincoln

Armston-Sheret, Edward. "Diversifying the Historical Geography of Exploration: Subaltern Body Work on British-Led Expeditions c.1850–1914." *Journal of Historical Geography* 80 (April 2023): 58–68. https://doi.org/10.1016/j.jhg.2023.02.004.

Armston-Sheret, Edward. "Tainted Bodies: Scurvy, Bad Food and the Reputation of the British National Antarctic Expedition, 1901–1904." *Journal of Historical Geography* 65 (July 1, 2019): 19–28. https://doi.org/10.1016/j.jhg.2019.05.006.

The University of Nebraska Press is part of a land-grant institution with campuses and programs on the past, present, and future homelands of the Pawnee, Ponca, Otoe-Missouria, Omaha, Dakota, Lakota, Kaw, Cheyenne, and Arapaho Peoples, as well as those of the relocated Ho-Chunk, Sac and Fox, and Iowa Peoples.

∞

Library of Congress Cataloging-in-Publication Data
Names: Armston-Sheret, Edward, author.
Title: On the backs of others: rethinking the history of British geographical exploration / Edward Armston-Sheret.
Description: Lincoln: University of Nebraska Press, 2024. | Includes bibliographical references and index.
Identifiers: LCCN 2024025254 (print)
LCCN 2024025255 (ebook)
ISBN 9781496230973 (hardback)
ISBN 9781496241337 (epub)
ISBN 9781496241344 (pdf)
Subjects: LCSH: Discoveries in geography—British—History—19th century. | Discoveries in geography—British—History—20th century. | Scientific expeditions—Great Britain—History—19th century. | Scientific expeditions—Great Britain—History—20th century. | Great Britain—Colonies—History—19th century. | Great Britain—Colonies—History—20th century. | BISAC: HISTORY / Europe / Great Britain / Victorian Era (1837–1901) | HISTORY / Expeditions & Discoveries
Classification: LCC G98 .A76 2024 (print) | LCC G98 (ebook) | DDC 910.941/09034—dc23/eng/20240715
LC record available at https://lccn.loc.gov/2024025254
LC ebook record available at https://lccn.loc.gov/2024025255

Set in Monotype Ehrhardt by A. Shahan.

For my family

CONTENTS

ILLUSTRATIONS

ACKNOWLEDGMENTS

Like the explorers discussed here, writing this book was made possible by the contributions of numerous other people. My deepest thanks must go to my wife and my best friend, Kiran Flynn. To list the innumerable ways that she has contributed to this project would be a book in itself. But I must say that I would never have even started this work (let alone finished it) without the influence of her infectious optimism.

I would like to offer my thanks to Innes M. Keighren and Klaus Dodds, who supervised the doctoral research on which this book is based. Both provided unfailing support, both during and after my time at Royal Holloway, University of London. Innes M. Keighren deserves special thanks for his generosity with his time, for his eagle eye, and for his consistent humor and encouragement. Discussions with and comments from Klaus Dodds have been invaluable in viewing the broader significance of this work. The detailed, supportive, and insightful feedback provided by my doctoral examiners, Dane Kennedy and Felix Driver, were invaluable in the process of writing this book and developing the arguments within it. I would also like to thank reviewers for a series of suggestions that materially improved this work. Bridget Barry at the University of Nebraska Press has also provided invaluable feedback and has been consistently supportive and encouraging of this project.

Peter R. Martin, Catherine Oliver, Sarah Pickman, Dane Kennedy, and Owen Vaughan deserve thanks for their encouragement and for generous feedback on parts of the text. Their comments and suggestions materially improved this book, and I am very grateful to them. I also owe a particular debt to my parents—Annie and Martin—for inspiring my interest in the history of exploration and encouraging me to think critically about these histories from a young age. They have also offered invaluable feedback on draft chapters of this book. I would also like to thank Jacob Henry for his help editing the introduction. While the feedback of others has dramatically improved this work, any errors that remain are, of course, my own.

Friends, including Andy and Dan Turner, Ben Newman, Adam Badger, James White, Henry Fry, Steve Lucas, Olivia Hough, and Ella Jeffreys have also been an invaluable source of support, as has my brother, Will. Meanwhile, I have valued conversations and encouragement from my colleagues in the Research and Higher Education Division at the the Royal Geographical Society with the Institute of British Geographers (RGS-IBG), including Catherine Souch, Sarah L. Evans, Phil Emmerson, Anna Lawrence, Julian Martin, and Emily Barker-Jubb.

I would also like to thank the staff at the archives of RGS-IBG, the Huntington Library, the National Library of Scotland, the Royal Asiatic Society, the American Geographical Society Library, the Royal Museums of Central Africa, and the Scott Polar Research Institute for their assistance in navigating these collections. Their insights on the archives drew my attention to several important items that I would have otherwise overlooked. Meanwhile, the team at the Techne Doctoral Training Partnership consistently supported my research and professional development.

Time and money are essential ingredients to any research project, and this book was also made possible by various forms of institutional backing. I was able to finish this book due to support from the Institute of Historical Research (IHR) through the Alan Pearsall Fellowship in Naval and Maritime History, which provided a supportive and welcoming academic community as well as time for writing, editing, and reflection. I would like to individually thank both Claire Langhamer and Jade Chesterton at the IHR for their work to make it such a positive and stimulating intellectual environment.

I would also like to thank the following organizations for grants that allowed me to carry out this research: the Arts and Humanities Research Council through the Techne Doctoral Training Partnership (Grant no. AH/L503940/1), the Royal Historical Society, the Historical Geography Research Group of the Royal Geographical Society, the British Society for the History of Science, the Adam Matthews Digital Archive, and the American Geographical Society Library at the University of Wisconsin, Milwaukee.

ON THE BACKS OF OTHERS

INTRODUCTION

Around 1850, high in the Canadian Arctic, a group of British naval explorers began to eat the bodies of their dead comrades.[1] They had set out in 1845 to find the Northwest Passage, a sea route between Britain and Asia, on two Royal Naval ships, HMS *Erebus* and HMS *Terror*. At first they were led by the British naval rear admiral Sir John Franklin, who hoped the steam-powered ships would find a way through the ice where previous vessels had failed. Now, with Franklin dead and their ships abandoned, the men resorted to the most desperate measures in their unsuccessful struggle to survive.

We do not know how the cannibalism started, whether they only ate the bodies of those who had already died, or if men were murdered to feed others. But we do know that the explorers started by dismembering bodies and scraping flesh from the corpses. As death neared, and the men grew more desperate, skeletal remains show that they probably cracked open human bones and boiled them to eat the marrow inside.[2] This extreme case of Arctic cannibalism exemplifies the central argument of this book: that expedition members came to rely on each other's bodies, and these relations of interdependence often involved behaviors shocking to domestic audiences.

The outside world first learned about this cannibalism from Inuit people who informed other British explorers like the Orcadian John Rae about the fate of Franklin's men. When Rae brought this news back, he was attacked in the British press, largely because his claims were based on Inuit testimony.[3] Charles Dickens dismissed the suggestion that Franklin's men had eaten each other, claiming that it was only based on the "vague babble of savages." He declared it "gigantically improbable" that educated, Christian, and disciplined naval men would resort to cannibalism, however hungry they were.[4] Rae and Dickens generally agreed about the trustworthiness of Indigenous people. Rae himself suggested that Inuit people would "not hesitate to tell a falsehood," and that the European explorer's job was to separate truth from fiction in such accounts.[5] Inuit people could not narrate their experiences to outside audiences without mediation—neither in Rae's account nor in the reports of later Victorian travelers who sought to investigate the issue.[6] The clash over the cannibalism question reveals how fundamentally Victorian prejudices have shaped exploration narratives and histories.

This book argues that we can develop a new and better history of exploration by focusing on the body. This might sound like a strange claim, as bodies are already omnipresent in adventure stories. We remember famous explorers for the trials and suffering they experienced as they undertook dangerous journeys, pushing themselves to the limits of endurance, and sometimes beyond. Explorers' bodies have sometimes assumed a quasi-religious significance. Objects associated with the Scottish missionary and explorer of Africa Dr. David Livingstone, or with Franklin, were referred to and displayed as "relics" in museums in Britain.[7] It is as if an item's proximity to the body of a now-dead explorer imbued it with some special power. Memorials to heroic travelers—such as paintings, photographs, and statues—also feature bodies. For instance, the headquarters of the Royal Geographical Society with the Institute of British Geographers (RGS-IBG) is flanked by statues of Livingstone and Sir Ernest Shackleton (fig. 1). Their bronze bodies display the different hardships these men experienced. Shackleton is dressed in a sledding outfit, complete with balaclava and thick mittens to guard against the frigid Antarctic cold. Livingstone's hands grasp his walking stick, and his belt is fastened on the tightest notch to emphasize the weight loss and weakness he suffered on his journeys. These heroic representations of explorers maintain the popular approach adopted by biographers, which is to laud remarkable individuals for their strength, character, and determination. Even today exploration is all too often framed by the language of individual heroism. This popular narrative overlooks the people and animals that explorers depended on and interacted with.

This book centers a different set of contributions and experiences of exploration and thereby challenges dominant narratives about the history of geography. Instead of focusing only on the bodies of expedition leaders, I devote attention to the multiple human and animal bodies involved in each journey, the various kinds of work they accomplished, and the ways

FIG. 1. Statues of Dr. David Livingstone and Sir Ernest Shackleton at the RGS-IBG. Author's photograph.

they were differently understood and represented across space. In doing so I build on the growing scholarship that has understood exploration as a "collective project of work."[8] I have been particularly inspired by Felix Driver, Lowri Jones, Dane Kennedy, Adrian Wisnicki, Lachlan Fleetwood, and Tiffany Shellam, who have examined the broader labor forces and relationships that made explorers' journeys possible.[9]

I develop this scholarship through my specific focus on the human and animal bodies involved in expeditions. In this sense I have been influenced by the growing academic interest in the role of animals in nineteenth-century science and empire, particularly the writings of Sujit Sivasundaram, Jonathan Saha, and Sandra Swart.[10] I also examine other ways in which bodies were important, including their roles in intercultural encounters. Explorers were themselves often the object of non-European curiosity, and expedition members engaged in sexual relationships—sometimes abusive, almost always exploitative—with women, girls, and sometimes men and boys they encountered.

This book is about *how* certain contributions and experiences were excluded from histories of the "geographical tradition."[11] In this sense it complements contemporary debates about the decolonization of geography—and the work of writers including Pat Noxolo, Tariq Jazeel, and Sarah A. Radcliffe—that have sought to challenge colonial legacies within the discipline.[12] As Ruth Craggs has argued, writing a history of geography that is attentive to a diverse range of contributions and experiences can play an important role in helping to understanding the challenges faced by geography today.[13] Many Victorian expeditions had more diverse labor forces than contemporary geography departments in Europe and North America. Yet diversity alone does not result in equal recognition. Interrogating how diverse expeditions became singularly associated with the names of white expedition leaders offers insights into how geography became colonial and how we might challenge this process. Confronting colonial legacies is an urgent task, made more apparent by the images and statues of colonial-era explorers that still adorn academic institutions.[14]

It is not enough to just "topple the racists" by removing these statues; we must also address the continuing legacies of exploration and empire in geography as a discipline.[15] In Britain academic geography remains an overwhelmingly white space, particularly at the professorial level, and the production of knowledge is still shaped by colonial legacies.[16] For instance, most academic fieldwork is still carried out by scholars from the Global North in formerly colonized countries. While scholars from the Global South must make complex, expensive, and often Kafkaesque visa applications to even hear research about their countries presented at academic conferences in Europe and North America. In the geosciences research in the field is often portrayed as a necessary "rite of passage" and is still linked to a rugged, outdoorsy vision of masculinity that contributes to the exclusion of minoritized groups.[17] "Going there" can yield important insights, but we must recognize that academic mobility remains shaped by the legacies of exploration and empire.[18] Studying the discipline's history—and exploration's important role in it—can help us better understand and dismantle such contemporary injustices.

It is therefore helpful to understand the changing ways that explorers and their bodies have been understood within the history of geography. For much of the twentieth century biographies narrating the lives of so-called great men dominated writing about explorers.[19] Reginald Coupland's mid-twentieth-century account of Livingstone's "last journey" is typical of these. Livingstone, Coupland argues, must be "ranked among the greatest of the great Victorians," and he attributes the explorer's achievements to his personal qualities.[20] While other kinds of literature have emerged, biographies of explorers remain popular—as a browse through the travel and exploration section of any commercial bookshop will show. The endurance of such modes of writing is not entirely surprising. After all, explorers make ideal and richly dramatic biographic subjects, as their lives are often full of adventure, hardship, and personal feuds. Not all biographical writing about explorers has been positive.[21] Cultural changes in the aftermath of World War II saw the emergence of more critical biographies, which often sought to recast once-celebrated figures as blunderers or villains.[22] While offering a different perspective from uncritical biographies of great men, these writings are often less radical than they first appear, as they remain firmly focused on individual explorers and their personal qualities. A more interesting approach to this subject has been "metabiography," which seeks to examine how biographies themselves have changed over time.[23]

While drawing on biographical writing where relevant, this book is inspired by scholarship that has sought to understand exploration as a social, political, and cultural phenomenon. Over the last forty years, an increasing number of writers have studied explorers as a group. Notable among these are postcolonial scholars, like Edward W. Said, Paul Carter, and Mary Louise Pratt, who have shown the important role that exploration and travel writing played within the ideological justification of European empires.[24] Explorers' accounts, they have argued, shaped the "imaginative geographies" that underpinned imperialism and "gave European reading publics a sense of ownership, entitlement and familiarity with respect to the distant parts of the world that were being explored, invaded, invested in, and colonized."[25] These works have been instrumental in shifting attention to the relationship between exploration and empire. But writers from within this camp, such as Frédéric Regard, primarily view explorers' texts as "*literary artefacts*" rather than historical sources.[26] They thus focus on issues of textuality and representation at the expense of the wider social, historical, and political forces (both domestic and in the field) that shaped explorers' journeys and the ways they represented them.[27] In viewing explorers as authors, we can lose sight of the fact that they were real people, with material bodies, who undertook journeys that could unfold in unexpected ways.

Scholars more familiar with historical methods have developed these works by drawing on detailed archival and contextual research. Notably, Felix Driver sought to understand exploration as a "set of cultural practices" involving "the mobilization of people and resources, especially publicity and authority."[28] Examining these "cultures of exploration" allows travel and fieldwork to be viewed in the context of broader developments, such as the expansion of European empires, shifting standards of scientific credibility and collection, as well as changing practices of writing and publishing.

One of the key issues within exploration cultures has been the question of credibility. Studies examining the relationship between travel and trust have become a particularly productive area of inquiry for historians of science and historical geographers.[29] As the historian of science Steven Shapin has demonstrated in *A Social History of Truth* (1994), domestic audiences had to make judgments about the truth of travelers' claims "at a distance," resulting in relationships of trust that became "inscribed in space." This meant that issues of trust and credibility were more acute for explorers than for other scientists, as the former made observations that were hard (or even impossible) for desk-based commentators to verify through replication or triangulation.[30]

Another challenge, noted by both Johannes Fabian and Dane Kennedy, was that many of the activities that effective explorers employed overseas, traveling in small parties and close (and sometimes sexual) relations with local people, rendered them problematic figures for domestic society unless their accounts were edited and sanitized.[31] Explorers also traveled in areas that pushed their bodies to the limit. For instance, Lachlan Fleetwood has shown that surveyors in the Himalaya often experienced altitude sickness in ways that affected their ability to produce accurate observations, a challenge that required both practical and rhetorical management.[32]

How experiences were narrated in books could also shape whether domestic audiences viewed explorers' findings as believable. As Innes M. Keighren, Charles W. J. Withers, and Bill Bell have shown, authors, editors, and publishers worked to ensure that travel narratives were viewed as trustworthy. While it is possible to identify common strategies used in different works, such "regimes of credibility" were "ad hoc," reflecting the individual judgments of those producing such works.[33] As they show, both bodily breakdown and the use of disguise could affect the perceived value of a traveler's observations and had to be managed "through careful judgment over different practices and strategies of self-presentation."[34] Another important issue is how changing publishing practices influenced exploration. What Beau Riffenburgh terms the "myth of the explorer" was shaped by the emergence of a sensationalist press in Britain and the United States. In sponsoring expeditions and publishing their findings, newspapers had a vested interest in elevating the explorer to the status of hero, often through drawing attention to the trials of their body.[35]

Bodies could also be helpful for validating explorers' findings. Disciplined, technical observations were widely lauded as the ideal quality of a scientific fieldworker.[36] But physical suffering could also be a source of authority.[37] As the historical geographer David N. Livingstone has demonstrated, the fact that "an explorer's body had undergone rigours and hardship in forbidding surroundings—literally bearing the marks of an alien environment—was considered the insignia of trustworthy testimony."[38] Dorinda Outram has also shown that domestic critics and readers "return repeatedly to the authentication of the explorer's travels by the trials of the body," when discussing the credibility of their findings.[39]

Academics have also addressed the role of instruments in the "regulation of observation."[40] Previous scholarship has shown that in the process of exploration, "objects and persons are mutually implicated in the production of knowledge in complex and intriguing ways."[41] Explorers employed a variety of narrative strategies to make credible knowledge, such as emphasizing their mastery of their own body and use of scientific instruments and photography.[42] Such strategies were often rendered problematic by the "simultaneous failures of instruments, inscriptions, and bodies."[43] Even so, attempting to use instruments—whether successfully or not—could be a source of credibility in its own right.[44]

Collecting botanical, ethnological, and zoological specimens would help render an explorer's narrative credible, and studies of collecting have been a valuable area of enquiry.[45] Lawrence Dritsas has shown that the scientific samples produced by an expedition could be distributed through global networks, relying on the labor of multiple people. Within such networks, the credibility of different samples was often established through relationships between "fieldworkers, institutions and 'analysts.'"[46] Given the importance of relationships in establishing what was true, it is unsurprising that collecting specimens did not always mollify domestic critics if they found wider issues with an explorers' assertions.[47] As Michael Heffernan has argued, explorers' claims were often "judged according to nonscientific criteria, notably the race, class, and gender of the person making the claim" and, sometimes, how they had dressed while traveling overseas.[48]

Within most literature on exploration, the focus is on the body of the expedition leader. This focus has been challenged by scholarship that has devoted attention to the "local partners, guides, porters, fixers, interpreters, traders, and officials who made journeys of exploration possible."[49] These insights have also reached broader audiences, through, for instance, Felix Driver and Lowri Jones's 2009 Hidden Histories of Exploration exhibition at the RGS-IBG's headquarters.[50] The exhibition showed how mainstream exploration histories often "overshadow the agency of these and other individuals."[51] Since then there has been a growing body of scholarship addressing the African, Asian, and Indigenous people who facilitated and participated in journeys led by Europeans. A vibrant literature exists on the important role of "indigenous intermediaries" and European leaders' dependence on preexisting skills and networks.[52]

Growing numbers of publications devote attention to investigating the broader labor forces involved in exploration—often through studies of specific expeditions or geographical regions—including particularly valuable recent contributions by Lachlan Fleetwood, Nanna Kaalund, Nokmedemla Lemtur, Peter R. Martin, and Adrian Wisnicki.[53] This book responds to and develops these works through focusing on the multiple bodies involved in expeditions and, in doing so, foregrounds the role of diverse groups of people within the history of geography.

This book also engages with the role of animal bodies in expeditionary parties. In part this multispecies focus arose out of the empirical material, which frequently discussed the animal contributions to expeditions. But it also reflects the growing academic interest in nonhuman bodies across the humanities and social sciences. The work of Dona Haraway has been particularly influential in critiquing the "anthropocentrism" of much Western research and writing and has questioned whether rigid distinctions between humans and nonhumans are ethical and accurate.[54] Historians, including Jonathan Saha and Sandra Swart, have carried out important work on animals and empire.[55] There has also been a growing interest in the more-than-human within academic geography as well as studies of expeditionary animals.[56]

The history of geography and the history of exploration, while closely related, are not identical. Sending expeditions to the blank spaces on European maps was one way to produce geographical knowledge, but it was not the only one. As David Lambert and Natalie Cox have illustrated, Britain had an illustrious tradition of "armchair geographers," who sought to map the world, not by traveling, but by reading the reports of others and synthesizing them with earlier classical writings.[57] Such methods fell out of favor in the latter part of the nineteenth century and were replaced by other efforts to define geography. From the 1880s several members of the Council of the RGS sought to direct the society's efforts away from exploration and toward the development of geography education and the establishment of the subject as a university discipline within Britain.[58]

Explorers often sought to portray their hard, muscular suffering in the field as distinct from, and superior to, sedentary geographers, who theorized from positions of comparative comfort. The body of active, adventurous, and manly explorers was often rhetorically contrasted with the elderly, flabby bodies of those who speculated from home.[59] Such efforts were also powerfully gendered. Mobility and credibility were intimately connected. For example, Lord Curzon's shifting views on admitting women to the RGS focused on what type of body a geographer required. In 1893 he opposed women's admission on the grounds that they were "unfitted" for exploration due to their "sex and training."[60] In 1913 he argued that women should be allowed to join, as "research work" would be of growing importance, and women, he claimed, were "as well equipped as men" for such activities.[61] Both his arguments for and against women's admission to the society assumed that exploration required a male body. The only thing that changed was the relative importance of adventurous expeditions within his understanding of geography's future.

We should not accept Curzon's claims about geography at face value. While there were sustained efforts to construct exploration in opposition to other branches of geography that involved less physical suffering, the practices were not always neatly divided. Explorers spent a great deal of time sitting and writing, both while composing diaries in the field and writing for publication at home. Explorers—such as Halford Mackinder, Douglas Freshfield, and Francis Galton—also became heavily involved in the promotion of geography education. Exploration has always had an impact on geography more broadly. Even today fieldwork remains a central part of geographical education and research, and organizations like the RGS-IBG continue to support expeditions through grants, training, and advice. Some academic geographers have sought to revive the expedition as a method of research, even while the climate crisis provokes renewed reflection on the centrality of travel-based geographical research and teaching.[62]

This book shows how diverse expeditions produced knowledge that overwhelmingly reflected white expedition leaders' perspectives and interests.[63] Ignoring or disparaging the embodied and collaborative labor of subaltern people was a powerful tool in making geography conform to dominant imperial ideologies. Moreover, I show how heroic understandings of exploration gloss over certain aspects of expedition leaders' bodies, particularly their leaky, lusty, and undisciplined qualities. A *colonial* geography focuses on the trials and contributions of the (almost always) white (and generally male) geographer, while ignoring the labor and contributions of the people they depended on and interacted with.

I argue that a geography that wants to do something different and better should recognize the underappreciated, embodied, and sensory work that contributed to producing and circulating the knowledge that sustains geographical institutions.[64] It must also acknowledge the important role that animals played in expeditions while remaining attentive to the ways in which animalizing language was used to propagate racial hierarchies.

Throughout this book I reflect on the relationship between exploration, embodiment, and empire. I challenge the mythology of explorers as "seeing men," whose powerful physique and commanding gaze embodied European domination.[65] In actuality explorers often depended on others and were even examined by them. Explorers had the "power to narrate" their experiences and cast themselves in more favorable terms, yet such rewritings never totally concealed the more complex realities.[66] Detailed archival research, contextualization, and critical analysis reveal the contributions of a far broader number of people and animals.

Many histories of exploration limit their scope to either one individual or one geographical region. While this allows writers to study the historical specificities of a particular area, it can obscure the similarities and differences between expeditions undertaken in different places. My comparative approach analyzes British-led expeditions to various areas, including Africa, Asia, South America, Australia, and the polar regions. I do not do so to show that exploration was the same everywhere. Instead, comparing these very different geographical contexts allows me to illustrate how expeditions were all shaped by their social, political, and environmental settings. My diverse selection of explorers and expeditions spans continents, environments, and periods (and also varying degrees of success, recognition, and fame). These examples demonstrate how contributions and experiences could be differently portrayed by explorers, depending on where they occurred and the identities of those involved.[67] Although the book's focus falls on British-led expeditions, and anglophone sources, the conceptual and empirical approaches I develop are also applicable to other national exploration cultures.

WHO WAS AN EXPLORER?

The scientific traveler was a nineteenth-century invention.[68] As Adriana Craciun has shown, the idea of the explorer as a specific cultural figure emerged due to changing practices of authorship and the establishment of geographical institutions like the RGS in 1830.[69] While there were important waves of travel and encounter before this period, the term "explorer" was not commonly used.[70] Nineteenth- and twentieth-century efforts to understand these earlier travelers as explorers functioned to "establish a genealogy of exploration that reinforced a sense of European exceptionalism."[71] Therefore, this study begins in the mid-nineteenth century and ends at World War I,[72] a symbolic and conceptually significant moment that changed the relationship between science, heroism, and empire in far-reaching ways.[73]

Exploration in this period was unique due to the variety of roles travelers performed and the cultural status they enjoyed. Max Jones has shown that explorers undertook a three-part mission: measure the world, measure manliness, and measure the relative power of different nation states and empires.[74] Every journey involved science or "discovery" (in some form) and an element of risk and danger.[75] While no central body governed who could—and who could not—be an explorer, institutions like the RGS tried to improve the quality of science carried out by travelers by issuing guidebooks, organizing training, and branding some fieldwork practices "scientific" and others "unscientific."[76] Yet the RGS could not give definitive instructions; they could only "hint" at how travelers should behave in the field.[77] Nor did such organizations have much control over who used the title "explorer." Many of the most famous explorers, like the Anglo-American explorer Henry Morton Stanley and Shackleton, had ambivalent relationships with the geographical establishment.

Even though no central authority ascribed "explorer" status, the term was shaped by Victorian and Edwardian prejudices about gentlemanly conduct.[78] The credibility of their findings was informed by broader cultural prejudices about who was (and was not) a reliable source of information.[79] Generally explorers were middle- and upper-class white men, though some did come from working-class backgrounds. African, Asian, and Indigenous people were seldom referred to as "explorers," "even when the regions they passed through were as unfamiliar to them as to the Europeans."[80] Any history of exploration must take care in labeling different people and different kinds of work. Terms that ostensibly celebrate Indigenous contributions, such as "guide" and "intermediary," often reinforce the inaccurate idea that such people did quantitatively or qualitatively different work than white explorers.

This leaves us with questions about how to refer to the diverse groups of people who made expeditions possible. In this book I use the word "subaltern," a term that has a history of its own: it was first popularized by the Italian Marxist Antonio Gramsci, who used it to refer to the broad array of classes—for instance, peasants, women, religious groups, and workers—oppressed by a given historical formation. In its literal sense, the term refers to individuals of "lower status" and has military connotations.[81] Gramsci used the concept to develop a Marxist analysis of colonialism, Italian history, and the relationship between the southern Italian peasantry and the working class.[82] Since then the word was popularized by the Subaltern Studies Collective, which sought to challenge both colonial and elite nationalist accounts of Indian history—initially by focusing on the "autonomous" agency and resistance of subaltern groups (particularly the peasantry) to British colonial rule.[83]

If used too broadly, the term "subaltern" can be unhelpful and homogenizing (e.g., on questions of gender difference).[84] As used here "the subaltern" does not refer to a specific group of people but a relationship to both power and representation.[85] This conceptualization is particularly helpful in the history of exploration, since the various subaltern people whom I discuss were not a social group in themselves. Instead they are unified by their subordinate position within a given expeditionary party and by their marginalization in the public and private records that explorers produced.[86] The term draws attention to the fact that it is often not possible to write a definitive "history from below" due to the limitations of the sources available.[87]

We must also reflect on the relationship between gender and exploration. The Victorian era saw a growing number of white women begin to journey across the world independently.[88] Historians disagree about whether they were "explorers." David Stoddart, for instance, justifies excluding women from the history of nineteenth-century geographical exploration since women (such as Isabella Bird and Mary Kingsley) who traveled the world and recorded their experiences in popular published accounts were "travelers," not geographical explorers.[89] Stoddart's distinction is based on the idea that women did not take measurements, make maps, or gather collections; they also rarely, if ever, referred to themselves as "explorers."[90] But many of these claims are simply inaccurate.[91] For instance, before traveling to Central Asia in 1890, Bird took courses in surveying and made celestial observations after her expedition's official surveyor went blind.[92] Alternatively many men became known as "explorers" even though their journeys contained relatively little science or discovery. By the early twentieth century, numerous white women from Europe and North America were undertaking journeys very similar to those made by men.[93] I analyze some of them in this book; if doing so blurs the boundaries between exploration and other forms of travel, then it succeeds in illustrating how arbitrary such classifications can be.

WHAT WAS AN EXPEDITION?

Explorers seldom traveled alone, so thinking in terms of expeditions, not individuals, can help account for all the people involved in making a particular journey possible. As Martyn Thomas has examined, expeditions were more than a collection of "individuals travelling in parallel."[94] A party was a "social unit, branded with its own sense of identity" in ways that shaped the experiences and encounters of each individual member.[95] But the blurry boundaries of expeditions can complicate this approach. They rarely arrived at their destination composed of the same people and animals who set out. For instance, Richard F. Burton recruited four separate groups of porters for his East African Expedition (1856–59) to Lake Tanganyika.[96] Each troop carried the expedition's supplies (and Burton himself) hundreds of miles across the continent. The animals used for labor, companionship, and food were often with the expedition for even shorter periods.

Expeditions also involved people and organizations who never left the relative comfort of metropolitan capitals: scientific societies, governments, politicians, and military officers instructed and supported explorers' journeys. In return travelers in the field sent letters back to armchair scientists and domestic institutions. As Dritsas argues, it is more helpful to think of any given expedition "not as a singular project or location, but as an aggregate of projects unified by goals, relationships, and responsibilities."[97] The porosity and complexity of expeditions is an important theme that I return to throughout this book. Focusing on expeditions (and the multiple bodies and relationships involved in them), rather than individual explorers, opens space for broader critiques of exploration and its relationship to power and empire.

Colonial-era travel narratives were both shaped by and constitutive of broader imperial ideologies.[98] Such narratives helped European readers understand and feel a sense of ownership over parts of the world they had never visited and, thus, laid the groundwork for conquest and colonization.[99] For instance, after his death, Livingstone became the "patron saint" of British imperialism in Africa, despite his own ambivalence about European empires.[100] Other explorers directly worked to establish colonial rule. Most infamously Stanley established trading stations that helped cement King Leopold II's influence along the Congo River. In doing so he contributed to the establishment of the racist, murderous, and exploitative Congo Free State, a political and commercial project that caused the deaths of around ten million African people between 1885 and 1908.[101]

While some explorers made land grabs for colonial powers, the relationships between travel, scientific institutions, and empire were not always straightforward. Geographical societies hosted vigorous debates about certain aspects of empire, and explorers did not always travel to places of interest to imperial powers.[102] Nor did travelers work exclusively for European empires—many mid-nineteenth-century explorers received support from African states (e.g., from Zanzibar and Egypt).[103] These nations' rulers hoped to expand their influence on the continent and used European travelers to establish their authority in its interior, particularly before the "Scramble for Africa" that followed the Berlin Conference of 1884–85. It is important to nuance the relationship between exploration and empire. But such caveats and complications should not distract from the fact that exploration *did* play a significant role in the expansion and ideological justification of European and North American empires.

It is no accident that most histories of exploration focus on the lives and experiences of a small group of white men (and sometimes white women). As scholarship in fields like subaltern and postcolonial studies has demonstrated, written and visual historical records reflect the unequal ways that information was collected, recorded, and preserved.[104] The historical sources explorers left for us reflect broader inequalities of power. Surviving published and private accounts were overwhelmingly written by expedition leaders and reflected the viewpoints of sponsoring institutions and publishers. Some leaders, including John Hanning Speke, even required expedition members to sign contracts that prohibited them from publishing their own accounts.[105]

Many records either ignore the experiences of racialized people or are overflowing with stereotypes and slurs. Sources written by subaltern people are far scarcer and, since most were transcribed by Europeans, are almost certainly also distorted.[106] Explorers' photographs and artworks raise different issues, but can sometimes offer glimpses of contributions and experiences invisible in written sources.[107] Expedition leaders deployed their "power to narrate,"[108] a power that ignored or distorted the non-European and working-class expedition members' perspectives and experiences (if their presence and contributions were acknowledged at all).[109] What Kathryn Barrett-Gaines terms the "unquestionable straightjacket of evidence" means that subaltern people are only "partially visible" in the archives.[110] Driver has shown that critical scholars must "decipher even the most unpromising of materials" and work "against the grain" of the archives to surface diverse voices and experiences.[111]

To accomplish this counterreading, I compare public and private materials to understand how certain people's contributions and experiences were distorted in published narratives. I often focus on incidents of debate and dispute within exploration cultures, as such moments most explicitly articulated broader assumptions. I also contextualize explorers' writings within the broader histories of racism and empire to situate the stereotypes and tropes that permeate their texts. Most importantly, my comparative approach juxtaposes how explorers differently narrated similar activities to reveal inconsistent assessments of relative value and importance, depending on where and by whom activities were undertaken.

Detailing expedition experiences beyond the European leader requires careful balancing around questions of responsibility. Foregrounding the work of subaltern people on British-led geographical expeditions demonstrates their essential role and complicates our understandings of what counts as skilled work. This helps unsettle the idea that "modern science" emerged in an "autarkic" Europe.[112] It is also important to emphasize that these individuals were not morally responsible for—nor had any control over—the expansion of empires. Exploratory journeys ultimately contributed to European empires' colonization of much of the world and the unequal transfer of knowledges and resources to colonial metropoles. Bringing this world into being involved a diverse array of people with differing relations to expedition leaders, colonial institutions, and state structures. It was also a process in which the body was an issue of central concern.

EXPEDITIONS AND BODIES

I center bodies and embodied labor to challenge the colonial narratives that permeate the history of exploration. An extensive academic literature has approached the body from a variety of perspectives and, in doing so, has raised broad epistemological questions.[113] As the historians Antoinette Burton and Kathleen Canning have argued, focusing on the body helps us think about, write about, and teach history in new and different ways. In particular it can foreground issues of gender and sexuality in historical analysis and allow us to study the lives, labor, and experiences of traditionally ignored subaltern groups.[114] Historians have also used a focus on the body to develop understandings of both racism and empire as well as resistance to them.[115]

Of course, the body is not a discrete object, separate from the mind. This assumed division—the fundamental precept of Cartesian dualism—has shaped dominant philosophical understandings of the human body in the post-Enlightenment West.[116] Such mind and body distinctions were (and still are) gendered, with women said to be constrained by their bodies and men being portrayed as more cerebral and rational.[117] This view was shaped by broader changes in understandings of sex and gender difference in this period.[118] During the nineteenth century, scientists in Europe and the United States also developed increasingly hardened views of racial difference, leading to the emergence of "scientific racism."[119] Such perspectives differed from earlier approaches, which emphasized the importance of language and environment in the emergence of human diversity. Instead many Victorian scientists claimed that there were innate, embodied differences between human populations. These binary distinctions diverged from how most people in the world understood the relationship between mind, body, and identity before the expansion of European empires.[120]

The term "body" raises deep philosophical questions about the relationship between bodies, materiality, and language.[121] Scholars including Michel Foucault and Judith Butler have shown the problems with viewing the body as a preexisting "biological given," since materiality, language, and society cannot be easily separated.[122] Such discussions are important, but their focus on language, power, and representation can sometimes draw attention away from "the body that eats, that works, that dies, that is afraid."[123] My approach addresses issues of culture and representation while acknowledging that bodies "remain material, physical, and biological entities."[124]

Exploration provides interesting terrain on which to examine such issues. Explorers' material bodies shaped their representations in profound and unexpected ways. An attack of fever, a fall, or an outbreak of scurvy at the wrong moment could leave indelible marks on a traveler's physical body, the kind of narrative they could produce, and its credibility to domestic readers.

It is worth emphasizing that this is a book about bodies rather than *the* body. I mean this in two senses. First, I am concerned not just with the singular body of the explorer but with the various human and animal bodies involved in expeditions and the complex and unequal relationships between them. Second, I am interested in how the bodies of singular people can be understood in different ways, as shown by Annemarie Mol.[125] Therefore, I analyze the same bodies using different interpretative frameworks in each chapter.

STRUCTURE OF THE BOOK

Chapter 1, "'Heroic' Bodies," challenges hagiographical biographers and postcolonial critics' readings of the explorer's body. Travelers' reputations were often contested, and conflicting representations of them—as heroes or villains—focused on different aspects of their bodies, such as their skin, hair, muscles, and forms of dress. Such discussions, I argue, reflected broader disagreements about what it meant to be an explorer, particularly the relative importance of science, heroism, adventure, and respectability, as each attribute relied on using and portraying the body differently. In chapter 2, "Dependent Bodies," I highlight how so-called explorer-heroes depended on other people's embodied labor. This included the muscular work of porters and chair-bearers, who carried supplies, instruments, and often the explorers themselves for hundreds of miles. Travelers also depended on the caring labor of individuals who nursed, cooked, and cleaned for them, freeing up time for scientific and observational work. I demonstrate the importance of this "body work" to the success of explorers' journeys.[126]

In Chapter 3, "Disciplined Bodies," I investigate questions of command and authority in explorers' published and private writings. I show how explorers would rhetorically manage their dependence on others by focusing on discipline. They frequently presented the success of an expedition as totally dependent on their control and framed subaltern expedition members as either dehumanized "instruments" of exploration or as unruly impediments.[127] I unravel these misleading descriptions and show that explorers often relied on the self-discipline of subaltern expedition members.

"Animal Bodies," the fourth chapter, examines the various roles that animals played in expeditions as laborers, companions, food, and zoological specimens. I also explore the challenges and opportunities of writing nonhumans into histories of exploration accounts. Writing animals into this history also draws attention to the interdependence of human and animal bodies and the contributions and skills of subaltern people in animal husbandry and the collection of zoological specimens. In addition, it sheds light on the ways in which descriptions of animals could be used to enforce and maintain racial hierarchies in published accounts.

In chapter 5, "Sensing Bodies," I examine the role of the senses in expeditionary knowledge production. In both biographical accounts and more recent academic literature, the explorer is often portrayed as a "seeing-man" who dominates landscapes with his disciplining and authoritative vision.[128] In contrast I show how explorers depended on the senses of (and oral information from) guides, intermediaries, and local people when making their (supposedly direct) observations. Antarctica presented a particular challenge for explorers, due to the lack of an indigenous population and the confusing array of optical illusions. Explorers often had to work together to understand what they were looking at and only gained confidence after spending prolonged periods in the continent.

"Examined Bodies," the sixth chapter, continues my critique of the literature on travel and vision by demonstrating that the production of knowledge was a two-way process involving multiple senses beyond vision. Explorers examined the bodies of others in ways that were exploitative and extractive. Yet, in the field, their own bodies were frequently subjected to scrutiny, debate, touch, measurement, and violence. Local people gained knowledge by examining British travelers. Even in Antarctica sailors' bodies were observed and examined by medics. Explorers frequently wrote about such encounters in ways that reasserted their control over the situation, exposing their own discomfort about being observed. These important incidents reveal the travelers' vulnerability in the field and prove that explorers were not the only ones generating new knowledge. Exploration was a two-way process of bodily encounter in which the traveler's body could be the object of others' curiosity. Recording these experiences in a certain way enabled explorers to reassert some control over the situation.

Finally, chapter 7, "Sexual Bodies," examines sex on expeditions to further dismantle the idea that explorers were detached, objective observers. Sexual and romantic encounters were an important aspect of intercultural encounters, but people within an expeditionary party also formed sexual relationships with each other. These relationships force us to abandon the idea that there was physical and emotional distance between explorers and the people they traveled with and encountered. Relationships were, in fact, embodied, emotional, and messy. This chapter also comments on the difficulties of studying such matters since explorers wrote about them in ways that would not damage their domestic reputations. Therefore, the question of sex on expeditions speaks to many of the fundamental problems facing historians of exploration.

Together these chapters reframe exploration away from its static heroic imaginary. In fact expeditions depended on numerous other bodies, who are seldom memorialized. Heroic representations of explorers' bodies impose a singular narrative onto these complex, hybrid, diverse journeys. This was not inevitable. Historical understandings of exploration were not monolithic, and travel did not automatically garner a heroic reputation. During their lifetimes many explorers had more complex public personas than we often realize. Such disagreements included detailed discussions of explorers' bodies and the ways they had been used at different times and in different spaces. We now turn to these issues and the contested relationships of explorers.

I

"HEROIC" BODIES

The statue of Sir John Franklin in the leafy shade of Waterloo Place, London, exemplifies the vision of exploration this chapter seeks to challenge. The memorial, erected by unanimous vote of the British Parliament, depicts Franklin, the "great Arctic navigator," in naval uniform. It was constructed in 1866 to memorialize his death while searching for the Northwest Passage. But this bronze figure and the plinth it stands on are not just a statue of an individual explorer. The monument also commemorates Franklin's "brave companions"—all 129 souls lost on the ill-fated expedition. Their bodies, shown only in a small bronze relief of Franklin's funeral, are not depicted in anything like the same detail as their leader's; their names are simply listed on the sides of the plinth on which Franklin's statue stands. One heroic body represents the entire expedition.

Absent, too, from this memorial is any discussion of why the expedition failed—including a scandal about the poor quality of the tinned food supplied to the explorers.[1] Nor is there any mention of less heroic aspects of the men's lives, such as the incidents of cannibalism that I began this book by discussing. Like most memorials to explorers, the emphasis is on the individual body of the expedition leader, which is portrayed in terms that gloss over the complexities inherent in explorers' journeys and reputations.

One important step toward developing a better understanding of expeditions is to reject the idea that explorers should be viewed through the lens of heroism. This is not an easy task. Victorian hagiographers tried to present the link between travel and courage as inevitable. "Polar enterprise seems to brace the explorer's morale, and converts him into a hero," one nineteenth-century Franklin biographer argued.[2] Such comments are far from aberrations. Contemporary scholarship often focuses on the "heroic reputations" of explorers, although they have done so for different ends.[3] Explorers feature prominently in recent works on heroism and empire, which emphasize how explorers embodied imperial values such as masculinity, progress, and conquest.[4] These ideas had purchase with the British public well into the twenty-first century, and several explorers—such as Ernest Shackleton, James Cook, Robert Falcon Scott, and David Livingstone—featured in a 2002 BBC poll to find who the British public thought were the "greatest British heroes."[5] All too often the explorers' body is understood as synonymous with the heroic body.[6]

In this chapter I argue that viewing explorers through the framework of heroism—often through a focus on the posthumous reputations of a narrow range of exemplary men—offers only a partial picture of the ways they were understood by their contemporaries. I offer a more subtle reading of the role of explorers' bodies in their reputations. I demonstrate that they were torn between competing ways of representing their own bodies that reflected underlying differences about what counted as acceptable conduct, ideas about gender, and the purpose of travel. Explorers faced a double bind: strength, discipline, and restraint were central tenets of adventurous heroism, helping to demonstrate their mental and physical prowess. But suffering, bodily breakdown, and adaptation to local circumstances were central features of travel narratives. Conflicting styles of expeditionary heroism and nonheroism were thus communicated, in different ways and at different times, through explorers' skin, muscles, nerves, and forms of dress.

Demonstrating the diverse response to explorers within the Victorian and Edwardian era is also important, given politically charged debates about statues and commemoration. Activists and academics, including the Rhodes Must Fall campaign, have asked why statues of British imperialists adorn prominent educational institutions around the world.[7] Recent scholarship has questioned whether explorers (many of them with troubling legacies) should continue to be displayed on the walls of the headquarters of the RGS-IBG at Kensington Gore.[8] Too frequently these debates are framed by the media as a "culture war" in which those questioning heroic interpretations of imperial history are presented as rewriting history.

In contrast I show that critical analysis of exploration is not a new development. Explorers were often fiercely criticized by their contemporaries, and many revisionist histories of exploration and biographies that have sought to cast explorers as villains have drawn on archival materials that show once-lauded explorers in a less favorable light. Often such criticisms were not made public (at least not in writing) in the Victorian and Edwardian era because it would have been considered ungentlemanly. In this sense modern reassessments of explorers' conduct and legacies are often a product of the fact that historians and biographers have access to information that was previously private. But even information made public at the time shows that explorers were understood in a variety of different ways.

These debates are hardly surprising when one begins to critically analyze heroism as a concept. The idea of the almost-invincible warrior hero has been an important feature of human cultures since ancient times. In the *Iliad*, for instance, Homer depicts Achilles as an "idealized male warrior" who is nearly "invincible."[9] Some argue that ideas of heroism have remained remarkably constant since ancient times. In *The Hero with a Thousand Faces* (1949), Joseph Campbell compares mythological and heroic tales from different regions and periods, showing that, throughout human history, these kinds of stories have shared common features and structures. Many explorers' tales fit quite neatly into the narrative structure of adventure, revelation, and return that, Campbell argues, characterizes the basic structure of the "monomyth" found within numerous world cultures.[10] Yet, in focusing on the similarities between heroic figures, Campbell's approach can lead us to ignore the unique roles heroes played in different contexts.

It is therefore important to situate notable individuals in their specific historical and social contexts.[11] The idea of the hero gained new importance in the Victorian era, due to a dramatic expansion in literacy, publishing, and the sensationalist press.[12] These changes led to the proliferation of growing amounts of writing that analyzed the personalities and achievements of "great men."[13] Such writing was linked to an individualist ideology that emphasized the importance of determination and "self-help."[14] Explorers' tales acquired a new significance because their journeys seemed to demonstrate the qualities of self-reliance and character that were becoming increasingly important at the time.

Moreover, heroism played an important role within British imperialism. The valorization of remarkable individuals—including explorers—contributed to the justification and popularization of European empires.[15]

Specific forms of "imperial heroism" emerged in Britain during the eighteenth century, which often lauded men who "displayed a range of exemplary qualities in the service of the nation on the imperial frontier."[16] Even when explorers were not directly promoting imperial expansion, they could still contribute to "the universalist rhetoric of scientific progress."[17] Victorians often preferred dead heroes, even if to modern eyes their deaths might be understood as avoidable, as the memorial to Franklin illustrates. In hindsight it is possible to see that the veneration of such "heroic failures" had important effects, making British imperialism appear, to certain audiences, as a "benevolent and just ruler rather than a tyrannical conqueror."[18] But what happens if we look at an example where the link between heroism and exploration was more complex?

CONTESTED REPUTATIONS: DRESS AND IDENTITY

The Victorian explorers Richard Burton and John Hanning Speke provide a different starting point for thinking about the relationship between bodies, heroism, and exploration. They were at the heart of one of the most dramatic disputes in mid-nineteenth-century geography: a debate about the source of the White Nile.[19] Within this dispute the two men's bodies, dress, and identity became the focal point of broader debates about conduct and credibility, as each tried to prove their respective geographical claims. Burton and Speke's expeditions took part in the growing European attention to the interior of the African continent and its hydrology. The interest in the source of the Nile not only reflected the importance of the river within biblical history but was also a product of growing European imperial involvement over the course of the nineteenth century, culminating in the "Scramble for Africa" in the 1880s.[20]

Burton was a prominent, if controversial, figure within Victorian geography. He first came to the attention of the British public after traveling to Mecca and Medina disguised as a Muslim pilgrim in 1853 and subsequently publishing an account of this journey. The venture made Burton famous, and he used this celebrity to gain support for an expedition to the Horn of Africa.[21] While preparing for this Somaliland expedition (1854–55) in Aden, Burton met Speke who then joined the venture.[22] Once in East Africa, the party of four split up: Burton proceeded in disguise, which was soon foiled, to the city of Harar (in modern-day Ethiopia). Meanwhile Speke tried to survey a geographical feature called the Wadi

Nogal.[23] After returning to the port of Berbera, the party were attacked by a group of Somali men, killing one of them (Lieutenant Stroyan) and leaving Burton and Speke with serious injuries.[24] The incident and its representation in Burton's book *First Footsteps in East Africa* (1856) annoyed and upset Speke.[25] Nevertheless he did not make his bad feeling clear to Burton at the time. Nor did this prevent Speke from taking part in another expedition led by Burton. The Royal Geographical Society (RGS) sponsored this venture, called officially the East African Expedition (EAE, 1856–59). The party left Zanzibar in 1857 to investigate the Great Lakes area of central Africa and its relationship to the Nile sources.

Burton and Speke traveled into the interior of Africa on existing trade routes, leading a diverse party composed of African and Asian soldiers, porters, and guides.[26] Despite suffering from fever, dysentery, and other tropical diseases, they became the first Europeans to reach Lake Tanganyika in 1858. Once at the lake, they could not hire a boat large enough to survey it properly and instead canoed around parts of it. Due to their own ill health and the hostility of the local population, they could not reach the northern end, where there was said to be a river called the Ruzizi, which flowed either into or out of the lake.[27] Unable to establish the direction of the river by observation and running low on time and trade goods, the explorers headed for home. On the way back, the pair briefly separated: Speke to investigate reports of another lake to the north, Burton to write up his anthropological observations and make plans for the return march.[28] On this journey Speke saw the body of water that he later named Lake Victoria and became convinced that it was the source of the Nile. Burton did not share Speke's conviction, leading to an increasingly public disagreement between the two men when they returned to England.

Central to this dispute was the relative credibility of the two men as witnesses. In Burton's case his enthusiasm for disguise and impersonation led to him being rendered as both a hero and a villain. European travelers had long adopted disguises to avoid detection and hostility when traveling in North Africa and the Middle East.[29] But they had also faced accusations that such misrepresentation was somehow less honorable and more duplicitous than wearing European garb. The apparent success of Burton's use of disguise on his first trip to Mecca and Medina brought him into the "national spotlight" among a domestic population fascinated by the Middle East and Islam and with a "fondness for cross-cultural role playing."[30] In his book Burton portrayed himself as an accomplished actor. He described how, in preparing for the expedition, he practiced the mannerisms of his alter ego, Abdullah, "to ensure they became second nature," changing the way he walked, sat, talked, and stroked his beard. He also underwent circumcision, although he only hinted at this in his published book.[31] Burton's use of disguise was greeted by many as a demonstration of his linguistic prowess and "his skills of impersonation."[32] Reviews praised his bravery, knowledge about Islam, and acting skills.[33] Throughout his life caricatures of Burton drew attention to his fondness for dressing in Arab clothes as well. But Burton's use of disguise proved controversial. In concealing his European identity, he opened himself up to accusations, particularly from evangelical Christians who objected to any form of impersonation.[34] Critics argued he was an untrustworthy source, as he had gained information by underhand means. Others suggested that his conversion to Islam might have been genuine.[35] By lying to gain information in the field, Burton undermined his credibility at home.

The idea that Burton was racially ambiguous and untrustworthy was a common refrain among those who met him. Even after removing Arab dress, some Victorian commentators did not view his body as fully British.[36] Richard's wife, Isabel Burton, claimed that his "brown," "weather beaten" skin and "Arab features" helped him disguise himself on his Arabian journey.[37] Alfred Bate Richards, who studied at Oxford alongside Burton, stated that he had "hands and feet of Oriental smallness," a "distinctly Arab look," and "piercing gipsy-looking eyes."[38] The poet Wilfred Scawen Blunt, who met Burton in Buenos Aires in later life, claimed that his face "lent itself" to disguise, "for it had in it little of the European, and there must certainly have been a cross in his blood, gipsy or other."[39]

A scar on Burton's left cheek, received when he was attacked in Somalia, gave him what one biographer describes as a "slightly sinister expression" (see fig. 2).[40] Windham Wyndham-Quin, the fourth Earl of Dunraven, even claimed that Burton "prided himself on looking like Satan—as, indeed, he did."[41] Some writers linked Burton's physical appearance to his credibility as a traveler. Blunt, for instance, claimed that Burton had the "eyes of a wild beast" before moving on to question the accuracy of his account of the journey to Mecca and Medina and repeating a rumor that he had murdered an Arab boy who had discovered his identity.[42] Burton's appearance was seldom referred to directly in the debates over the Nile's sources. But many within the RGS certainly viewed him as something of a "liability," which shaped their decision to send Speke—rather than Burton—back to East Africa to carry out further investigations.[43]

Burton's defenders felt compelled to emphasize that his body was visibly English. His niece and biographer, Georgiana Stisted, complained, in racially offensive terms, that most portraits of Burton were "simply hideous; the skin is the colour of a brown monkey's, the features coarsened and exaggerated." She claimed indignantly that "with Burton's marked look of race, he never could have been taken, unless purposely disguised, for other than an English gentleman."[44] Like the explorer's critics, Stisted saw the suggestion that he was able to easily disguise himself as an Asian man as an insulting accusation that threw into question his status as a gentleman.

The final resting place of Burton's body shows how after his death he remained an eccentric figure. He is buried in a sandstone mausoleum in the shape of a Bedouin tent in the graveyard of the St. Mary Magdalen Catholic Church in Mortlake, London. The remarkable structure, designed by Isabel Burton, who rests alongside him, features both a prominent crucifix and numerous crescent moon and star symbols (often used to represent Islam). Even the location of the memorial in a Catholic cemetery seems incongruous for such a lifelong critic of established Christian values (and was only made possible by a deathbed conversion to his wife's faith). Even after his death, Burton's body brings to the surface questions about who he really was and where his cultural loyalties lay.

Both in body and character, Speke was very different from Burton. His "physique was emphatically Anglo-Saxon," and he was apparently "tall and handsome in the English style, with blue eyes and a brown beard."[45] Burton described Speke as "a man of lithe, spare form, about six feet tall, 'blue-eyed, tawny-maned; the old Scandinavian type, full of energy and life.'"[46] His body and conduct made him a less racially and morally ambiguous figure than Burton, and it was he who was sent back to East Africa on a further RGS-supported expedition with another army officer, James Augustus Grant, in 1860. The two men succeeded in traveling around Lake Victoria and managed to follow the Nile for much of its length, eventually reaching Cairo in early 1863. Despite collecting

FIG. 2. An undated photograph of Richard Burton, showing the scar on his cheek. Courtesy of the Wellcome Collection. Image ref: 12443i.

further evidence, the expedition failed to convince Speke's critics—and even some supporters—that Lake Victoria was the source of the Nile. The pair had been forced to depart from the lake for considerable periods (meaning they could have seen several unconnected different bodies of water), and there were issues with the quality of their observations.

Unable to establish the source of the Nile through evidence and observation alone, Speke tried to use representations of his body to emphasize the relative credibility of his claims compared to Burton's. In *Journal of the Discovery of the Source of the Nile* (1863), Speke included an image of himself standing before Lake Victoria as the frontispiece to the first edition of the book (fig. 3). In the portrait he is dressed in distinctly European (if practical) clothing. The caption emphasized that the picture showed Speke in the "dress worn in Africa."[47] The image drew attention to his identity as a British gentleman and thus his credibility as a witness in the eyes of British commentators. Speke elsewhere claimed to have a moral objection to disguise.[48] He described how on the EAE his guides suggested he wear Arab clothing "in order to attract less attention."[49] Speke questioned their motives and claimed that he was "more comfortable . . . both mentally and physical" in European clothing than if he had "degraded" himself by wearing "their hot, long, and particularly uncomfortable gown."[50] These passages are clearly an implicit attack on Burton, aimed at subtly undermining his credibility through reference to the way he had dressed.

Speke also sought to distance himself from Burton by emphasizing how little he had adapted to local customs. In *Journal of the Discovery of the Source of the Nile*, he described an incident where he was asked to sit on the floor in front of Kabaka Mutesa I, ruler of Buganda: "I had made up my mind never to sit upon the ground as the natives and Arabs are obliged to do, nor to make my obeisance in any other manner than is customary in England."[51] Speke's book included an illustration that showed him and Grant sitting on stools next to the kabaka, while other men sit on the ground. In demonstrating his conformity to British norms, Speke was again distancing himself from Burton, the enthusiast for disguise and impersonation.

But to justify his inflexibility to local rulers, Speke lied. He pretended to be a British prince to African leaders in the hope that it would secure him better treatment and a direct audience with African royals. On arrival in Buganda, Speke was offered some huts used by Arab traders to stay in, but he protested, claiming that he was a "foreign prince, whose royal blood could not stand such an indignity."[52] He continued to portray himself as a man of "high rank and importance in his own country" for the remainder of his journey.[53] Speke's deception earned him criticism, particularly from the armchair geographer James MacQueen, who used it to question his geographical credibility in a withering attack, published in collaboration with Burton.[54] Even though Speke portrayed his body in ways that conformed to the domestic prejudices, he was not always viewed as a hero.

FIG. 3. Portrait of John Hanning Speke in dress worn in Africa, adapted from a photograph by the Southwell Brothers as used in *Journal of the Discovery of the Source of the Nile*. Courtesy of the New York Public Library. Image ref: 1258377.

Burton and Speke's public reputations incorporated issues of dress, identity, and adaptation in different ways. Burton emphasized his cultural knowledge, linguistic skills, and ability to disguise himself; Speke, by contrast, stressed a specifically English version of expeditionary heroism, based on a maintenance of most domestic norms when it came to dress and conduct. Both versions were not universally successful and each attracted criticism in some quarters. Neither explorers' body can be understood as unambiguously heroic.

The role of dress in an explorers' reputation was even more acute for women travelers. Dominant cultural norms discouraged women from participating in dangerous journeys to far-flung places, particularly if unaccompanied by husbands.[55] But women did challenge these norms. For instance, Isabella Bird journeyed to every continent except for Antarctica between 1872 and 1904 and published numerous books and articles on her travels. Moreover, she holds a critical position within British geography and was one of the first women admitted to the RGS in November 1892. Like other women travelers, Bird faced an additional pressure to maintain domestic standards of modesty and decency when overseas. On most of her journeys, Bird wore a tweed dress, which she referred to as "the Hawaiian riding dress": "a half-fitting jacket, a skirt reaching to the ankles, and full Turkish trousers gathered into frills falling over the boots."[56] Bird emphasized that the outfit allowed her to maintain both her modesty and femininity. In the first edition of her breakthrough work, *A Lady's Life in the Rocky Mountains* (1879), Bird wrote that she wore her "Hawaiian riding dress over a silk skirt, and a dust cloak over all."[57]

Bird grew "annoyed" when one reviewer claimed that she had "donned masculine habiliments."[58] Another review similarly claimed that Bird had adopted a "male riding attire," and that she was "at first rather shy of appearing in it."[59] Bird saw such descriptions as a direct attack on her public reputation because of wider expectations about middle-class women's modesty at the time.[60] Writing to her publisher, Bird claimed that the accusation was "simply horrible and a lady ought to be protected from such truly 'libellous' treatment on the part of reviewers."[61] The Hawaiian riding dress was, she insisted, "the dress that is worn by ladies at mountain resorts in America and by English and American ladies on Hawaii," and she noted that one of her friends had claimed that "a more feminine dress for rough travelling could not be devised."[62] In most future editions of the book, she included a drawing of herself in the dress and ensured that a note was included to explain her outfit, describing it as "a thoroughly serviceable and feminine costume for mountaineering and other rough travelling."[63] The entire incident illustrates how careful women travelers had to be when describing their dress in order to avoid domestic criticism.

Other women grappled with similar challenges. Mary Kingsley, who explored West Africa in the 1890s, struggled to dress in a way that was both practical and socially acceptable. Kingsley felt this challenge particularly acutely in the presence of Europeans from other colonizing powers. She tried to keep one dress in a relatively good state for such occasions, but complained that a West African traveler's "wardrobe is always in a rarefied state."[64] As with Bird, Kingsley sought to manage gendered expectations about her dress through the ways she wrote about her journey. In her books she used humor when describing her somewhat bedraggled appearance when meeting dignitaries.[65] Through self-deprecation she demonstrated her respect for European social norms, even as she transgressed them—though some reviewers complained about her "slangy" tone.[66] Upon her death obituaries of Kingsley emphasized that, despite her adventurous travels, she had maintained her femininity. The *Pall Mall Gazette* described her as "rather under the medium height, slight and somewhat spare . . . fair in complexion and colouring. Her beautiful deep blue eyes were the most arrestive feature of her face." The review also emphasized that she was always "neatly and unobtrusively dressed" in "ordinary costume"; such attire, it noted, separated Kingsley from the "new women" of the late nineteenth century who sought to question established gender and sexual mores, for instance, through dressing in "rational" clothing.[67] These comments illustrate the degree to which positive evaluations of women's travels often remained contingent upon them using and representing their bodies in very specific ways—even if changing social norms meant that women explorers were more likely to receive a positive response than twenty years earlier.

Women might still use expeditions to dress in new ways. When Bird traveled to China in the 1890s, she wore an adapted version of Manchu women's dress (fig. 4).[68] She described "the extreme comfort of a Chinese woman's dress in all classes, no corsets or waist-bands, or constraints of any kind."[69] Bird returned from China with samples of Manchu dress and included a copy of this image in *The Yangtze Valley and Beyond* (1899), using it to emphasize her worldliness and appealing to a growing British interest in Chinese culture.[70] Bird, unlike Burton, never pretended to be Chinese, simply dressing this way to attract less hostility. Even so she publicly claimed that she "never became reconciled" to one mode of Chinese dress that included trousers without a covering skirt or petticoat.[71] Warned by her earlier experiences, Bird sought to emphasize that her embrace of Chinese clothing had not led her to transgress certain

FIG. 4. Isabella Bird (Mrs. Bishop) in Manchu dress. Courtesy of the New York Public Library. Image ref: 826842.

domestic social norms. She also found Chinese dress particularly comfortable because she had a long-standing spinal condition that caused her extreme pain for long periods. This condition only seemed to get better when Bird was traveling.[72] In contrast to other travelers, for whom the field was a site of suffering, for Bird it was a place that allowed her to dress more comfortably and to recover her health. Focusing on the issue of suffering can shed further light on the role of the body in Victorian and Edwardian exploration cultures.

WEAKNESS, NERVES, AND GUTS

Bodily suffering was an essential feature of exploration accounts. But it was a fraught area as well. The ambiguous role of suffering in explorers' accounts is illustrated by returning to Burton and Speke's public disagreement about the sources of the Nile. This debate led each explorer to portray the other as weaker and less manly than the other. Burton and

Speke freely admitted that they had suffered severely on the East African Expedition but did so in different ways. Speke portrayed himself as tough and adventurous. In *What Led to the Discovery of the Source of the Nile* (1864), Speke reported a gory incident of a beetle crawling into his ear and his unsuccessful attempts to remove it with tobacco, salt, and, eventually, a knife. Speke claimed the incident was "the most painful thing I ever remember to have endured," yet stated coolly that it was "not altogether an unmixed evil," because it apparently drew inflammation away from his weakened eyes.[73] In such passages he presented himself as able to withstand extreme suffering. Speke also described the "marvellously rapid manner" in which his severe wounds, sustained during the Berbera attack, healed before the start of the EAE. He claimed the recovery was mostly a result of his "strong constitution," and the fact that he had been "living for months in a very abstemious manner."[74] He therefore presented his physical strength as a product of his ability to discipline his appetite—an increasingly popular belief in the Victorian period.[75] Drawing on such descriptions, later explorers described Speke as a "man of action" who had a "fine figure" and a "martial bearing."[76]

Burton, in contrast, drew on religiously inspired ideas of sacrifice, stressing the physical price he had paid for the expedition's geographical discoveries.[77] He publicly claimed that fever nearly killed him, preventing him "from walking to any distance for nearly a year."[78] Isabel Burton adopted a similar approach in her biography of her husband. She described how, on returning from East Africa, Richard "was a mere skeleton, with brown yellow skin hanging in bags, his eye protruding, his lips drawn away from his teeth."[79] In highlighting his sufferings, such descriptions spoke to powerful preexisting narratives of "sacrifice and martyrdom" popular in Britain in the nineteenth century.[80] But in Burton's case they were not so successful, because of the public conflict with Speke.

Both men sought to portray the other as weaker than themselves. Speke claimed that Burton "complained about the shock his nerves had received" during the attack in Somalia and suggested this affected his performance and bravery for their journey.[81] Speke also stated that Burton had suffered more severely from fever.[82] Burton's annotations in the margins of his personal copy of Speke's book demonstrate his annoyance with such allegations.[83] Beside one passage describing how

Burton suffered severely from fever, Burton jotted, "He was worse."[84] Such comments illustrate the petty nature of the dispute between the two men. But, more broadly, they show that Burton, like other explorers, viewed descriptions of his body as an issue worthy of comment and correction. Burton publicly portrayed the two men's strength in terms diametrically opposed to Speke's. In *The Lake Regions of Central Africa* (1860), he disparagingly referred to Speke as his "invalid companion" and suggested that he required more assistance on the march.[85] Burton claimed that during the ascent of one mountain pass they were enfeebled by fever: "My companion was so weak that he required the aid of two or three supporters; I, much less unnerved, managed with one."[86] These passages show the competing pressures explorers felt around the issue of suffering, for both men wanted to emphasize the hardship they experienced, but neither wanted to be seen as less hardy than the other.[87]

Women explorers faced slightly different challenges when it came to such matters. An emphasis of physical weakness and frailty could be an important way for women travelers to portray their femininity. In *A Lady's Life in the Rocky Mountains*, Bird described her ascent of Long's Peak in Colorado. She claimed that a local man named Jim Nugent, "dragged" her up "like a bale of goods, by sheer force of muscle," and she afterward lamented that her companions must have been "grievously disappointed" in her "courage and strength."[88] Bird was more willing to include such descriptions of dependance than male explorers. But some reviewers had little sympathy for her. A reviewer of the book claimed that Bird traveled "for the sake of variety of excitement" and that she therefore "forfeits all pity" for the suffering she experienced.[89] Such criticisms were implicitly based on the idea that adventurous travel was a masculine preserve, and that Bird was ill advised in undertaking such a journey. Bird's travel accounts thus worked "both with and against the hegemonic ideal of genteel-class British chauvinism."[90]

In her later travels, Bird downplayed the dangers she experienced. This approach made clear to readers that she "had faced dangers and difficulties from which a man of nerve and courage might have shrunk."[91] For instance, she continued traveling through China even after being attacked by a large and hostile crowd, describing the incident in an offhand manner.[92] But the suppression of suffering to appear heroic could also lead to problems.[93] When one reviewer of *Korea and Her Neighbours*

(1898) praised Bird for her "cool saneness," she told her publisher John Murray: "I think I have perhaps too much minimised in writing on Korea and Western China the very serious discomforts and difficulties as well as the actual risk which I ran."[94] Like other travelers Bird struggled to write about the pain she experienced in an appropriate way.

Bird also emphasized the fact that she traveled to try and improve her health.[95] At home she often suffered from prostrating chronic health conditions. While overseas she was capable of incredible feats of endurance, suggesting that her ill health was, at least in part, the product of social and psychological factors. Victorian and Edwardian commentators frequently struggled to reconcile such conflicting descriptions and searched for an explanation. The idea that Bird disciplined her body through acts of mental willpower was a common theme in contemporary commentary on her expeditions. An Edwardian biography noted that "the minor worries of housekeeping assailed and depressed her, while danger and difficulty appealed to her marvellous self-control and resource."[96] The *Geographical Journal*, meanwhile, claimed that Bird's journeys represented "the triumph of human will over the effects of disease and bodily weakness."[97] Similarly, an Edwardian book on nineteenth-century women travelers praised Bird for "her true patience, her unfailing good humour, and her indomitable 'pluck.'"[98] Bird's weak body rendered her achievements more impressive and was used to illustrate her strength of character, bravery, and self-control.

Bird's moral strength could even be mapped onto parts of her body. Following her death in 1904, an obituary of Bird in the *Edinburgh Medical Journal* suggested that her health was "very much a paradox."[99] The journal explained such discrepancies, suggesting that Bird was "one of those subjects who are dependent to the last degree upon their environment to bring out their possibilities."[100] Climate and situation seemed to play an important role in her capacity for strength and adventure overseas. But the *Edinburgh Medical Journal* also claimed that, in order to travel "some 2400 miles in wild and untamed country," Bird must have had "a basis of strength of an unusual kind" and a "large storehouse of energy behind it all," but gave little clue as to what its source might be.[101] Her Edwardian biographer Anna Stoddart offered some ideas about how Bird was able to overcome "the serious ailments which exhausted her constitution." She explained that "her appetite and her power to assimilate large quantities of food healthily never failed. . . . She could go on for days with little more than a bowlful of rice and a handful of dates and raisins; but when substantial food was to be had she availed herself amply of the opportunity. Indeed, she called herself a 'savage' in the matter of food."[102] Strong guts seemed to enable Bird to overcome her bodily weakness.

While commentators were fascinated by Bird's complex constitution, they expected British naval explorers to conduct themselves in more stereotypically masculine ways. From the early nineteenth century, frozen spaces were constructed as places where the "disciplined, resourceful British naval men almost succumb to starvation and madness," and where "the resistant, intact, male body so central to imperial heroic masculinity meets its limits."[103] But what happened if the naval body failed in non-heroic or avoidable ways? Studying the British Arctic Expedition (BAE, 1875–76) allows us to examine this dilemma further. Led by Admiral George Nares, the large, well-equipped naval expedition set out to try and reach the North Pole via Baffin Bay on two ships, HMS *Alert* and HMS *Discovery*. Nares hoped to find an "Open Polar Sea" that would allow the explorers to sail to the North Pole. Instead they found impenetrable ice. Nares dispatched a party, led by Commander Albert Hastings Markham, that sought to sledge to the Pole. While Markham's group managed to set a new farthest north record, they turned back farther south than hoped. Worse still the expedition suffered severe outbreaks of the "terrible disease" of scurvy, affecting men on board the two ships and those sledging.[104] Concerned by the death of several men and the illness of many others, Nares ordered the two ships to return home after only one winter in the ice.

When the expedition returned to the UK, after an initial flurry of enthusiasm there was soon widespread dismay that the expedition had spent less than two years in the Arctic and that four men had died. As one member of Parliament commented, "No one had expected the Expedition to return within one year without having accomplished its object."[105] The medical press complained about the fact that a rum ration had been issued to the explorers.[106] Soon the Admiralty was forced to hold an inquiry into the issue, as the press, medical journals, and MPs criticized the organization of the expedition and the way the men's bodies had been cared for. In private correspondence with the secretary of the American Geographical Society (AGS), the Arctic explorer John Rae claimed "that had not the Admiralty been in such great haste in bestowing rewards and honors, a court martial would have been the result."[107] In its final report, the Admiralty blamed the scurvy outbreak on the diet of the explorers, while suggesting that there were problems with dampness and bad air on board the ships. Blame was additionally attached to individual bodies and behaviors. The committee claimed that many of those worst affected by scurvy had been "addicted to the immoderate use of alcohol."[108] Their suffering had been avoidable and was therefore not heroic.

Some drew broader conclusions about the lessons to be drawn from this failure. In a parliamentary debate on the expedition, the Liberal MP Lyon Playfair lamented the scurvy outbreak, linking it to a broader problem of military scurvy experienced in recent conflicts. He saw studying the expedition as central to averting further disasters, commenting that "the lessons to be derived from that Expedition are vastly important to the welfare, not only of our naval, but also of our land Forces, in many conditions when limited food is accompanied with heavy labour."[109] For him the failure of the BAE thus highlighted the need to strengthen imperial defenses by better caring for the British bodies at the edges of the empire. Far from being portrayed as heroic, masculine, and strong, the explorers' bodies triggered broader anxieties.

Such anxieties became more acute in the last decades of the nineteenth century. By the early twentieth century, dominant visions of masculinity increasingly focused on the virtues of physical strength and self-discipline; this led to concerns that modern, sedentary urban life was "degenerating" men's bodies and rendering them soft, flabby, and feminized.[110] Rugged, outdoorsy masculinity became an important feature of exploration cultures.[111] Concerns about the vitality of British bodies became particularly acute in Britain in the run-up to World War I, partly due to recruitment scandals during the Second Boer War (1899–1902) in which large numbers of British men were deemed to unfit to serve.[112] Polar exploration was increasingly a way of proving the continuing vitality of the British body, and incidents of suffering and vulnerability that featured prominently in earlier travel accounts now went "unspoken."[113] But this new culture presented problems when explorers' bodily performances did not conform to these ideals.

The challenges faced by prewar explorers are illustrated by press discussions around another scurvy outbreak, this time the British National

Antarctic Expedition (BNAE, 1901–4).[114] The expedition was led by Commander Robert Falcon Scott, organized by the Royal Society and RGS along naval lines, and received government funding. The expedition traveled south on a specially built ship named the *Discovery*. While in Antarctica the expedition experienced numerous problems with its tinned food supplies and, like the BAE, an outbreak of scurvy. The disease first manifested itself on sledging trips in October 1902, but signs of scurvy were found in many more men on board the ship. Robert F. Scott, Edward Wilson, and Shackleton also suffered a severe outbreak of the disease on their farthest south journey in 1903, which affected Shackleton particularly severely. The outbreak was dealt with without fatality, and, when news reached London, much press reporting of the expedition presented scurvy as an inevitable peril of "Antarctic adventures."[115] But some viewed the outbreak as a national scandal, suggesting the disease raised broader questions about British institutions.

The critics' major gripe was that the expedition had received significant government funding, and that other countries were able to run polar expeditions without such outbreaks. The *British Medical Journal* was particularly scathing, lamenting the fact that the expedition had been "crippled" by the disease.[116] The author claimed that "some one has blundered" in the equipping of the expedition and suggested that the problem lay in the BNAE's naval organization.[117] The author noted the "somewhat humiliating contrast" between British and Norwegian expeditions, the latter of which had spent much longer periods in the polar regions without succumbing to the disease.[118] The *Daily Mail* was even more critical, claiming—inaccurately, in the case of the British Arctic explorer Frederick Jackson—that "neither Nansen on the sea nor Jackson on the land had a single case of scurvy. Yet the *Discovery*, in the very first year of her absence is visited by the scourge."[119] The *Daily Mail* argued that the root of the problem lay in the "bickering" between the RGS and RS, which distracted them from their duties.[120] As with the BAE, scurvy seemed to demonstrate the failure of bureaucratic and divided institutions to care for the bodies of those they sent across the world.

The speeches of RGS president Sir Clements Markham only fanned concerns about the safety of the BNAE. In 1903, around the same time as news of scurvy broke, Markham was trying to muster support for a relief expedition to resupply the *Discovery*. To raise funds he claimed that the explorers were in mortal danger if supplies were not sent. Speaking to the RGS, Markham claimed that without further funding "those heroes who have done so much for science and their country's credit will be in grave peril."[121] Although such a narrative, which stressed the explorers' sufferings in the service of science, proved effective at securing government support, it also presented the expedition as weak and emasculated.[122] The *Standard*, for instance, expressed "some anxiety about the safety of the expedition" following news it would spend another winter in Antarctica.[123] The *Speaker* claimed that while the scurvy outbreak "was not serious, and yielded readily to treatment . . . the expedition would certainly have starved to death" if it were not for the relief ship and availability of seal meat.[124] The *Speaker*, like others, saw the near disaster as part of a wider problem, commenting, "There have been too many scandals of the kind in connection with our Polar expeditions."[125] When the relief expedition eventually reached Scott in 1904, the *Evening Telegraph* claimed that the *Discovery* had been "recused" from Antarctica.[126] Such passages show that the idea that something had gone wrong with the organization of the BNAE was a significant (if not dominant) feature of press reporting.[127] These accounts portrayed the expedition's bodies in terms very different from prevalent ideas of masculine heroism in the polar regions. In his book Scott sought to reframe the expedition's experiences in more conventionally heroic terms.[128] He also sought to rebuke the idea that the expedition had needed rescuing in his book, suggesting that life in Antarctica was "very pleasant."[129] While he had some success in recasting the expedition's experiences, and his book was positively reviewed, the bodies of the explorers were not solely understood as heroic.[130]

DEAD BODIES

If explorers faced criticism while they were alive, was death the point at which they became heroes? Many of the most widely lauded travelers were those with established reputations that died on their journeys: Robert O'Hara Burke and William John Wills, Sir John Franklin, David Livingstone, and Robert Falcon Scott. The importance that Scott's death played in establishing his reputation as a hero was even commented on at the time. After Scott's body had been discovered, the expedition's second-in-command, Captain Edward Evans, wrote to the naval officer and polar explorer Albert Markham expressing his expectation that descriptions of the explorers' suffering would squash critics of the expedition. He hoped that once details of the men's last days were made public "those 'rotters' who wrote describing Scott's life work as a 'picnic to the Antarctic' will slink away ashamed."[131] Evans's quote demonstrates that even contemporary writers understood death as a key moment in developing an explorers' heroic reputation. He also implies that some saw polar exploration as a "picnic."

The heroic image of the explorers that emerged centered on certain aspects of their bodies. In early November 1912, a search party found the frozen corpses of Scott, Wilson, and Henry "Birdie" Bowers on the Ross Ice Shelf, Antarctica. After cracking Scott's arm to remove the journal tucked beneath it and collecting a variety of papers and relics, the search party decided to bury the explorers' bodies where they had died.[132] At the memorial service held in St. Paul's Cathedral on the February 14, 1913, the explorers' material, fleshy bodies—thin, sunburned, frostbitten, and scarred by the effects of polar travel—were absent. This absence was partly a product of the impracticality of moving the explorers' cadavers all the way from Antarctica, and partly because the sublime polar landscape was viewed an appropriate resting place. Whatever their motivations, by leaving the bodies in Antarctica, the search party ended up shaping the way the expedition was remembered. With their material bodies absent, a mythical picture of the explorers emerged. Scott's final writings and Captain Oates's self-sacrifice were "singled out for praise."[133] In popular culture Scott's polar journey was transformed into a quest in which their suffering demonstrated their moral qualities of determination and perseverance.

Following Scott's death writers emphasized his moral rather than physical strength. To many the way Scott and his companions died offered "a reassuring example of heroic character and idealism."[134] In his postwar analysis, Apsley Cherry-Garrard, the assistant zoologist on the expedition, claimed that his experience had convinced him that being a polar explorer was "a matter of mind rather than of body."[135] Cherry-Garrard's analysis is reflected in scholarship on the expedition, which remains focused on the disembodied and moral aspects of Scott's heroism. American cultural critic Lisa Bloom argues that the focus on character in the aftermath of Scott's death fed into "an ideology of masculinity in which paradoxically the male body is ignored," and suffering demonstrates the triumph of character and control.[136] But we should not overlook the important role that Scott's body played in his heroic reputation.

Before he died descriptions often focused on his physical prowess. For example, in late 1910, Joseph James Kinsey, the expedition's representative in New Zealand, gave a press interview in which he claimed that Scott was the ideal person to reach the pole because he had "the physique and the determination, and he can endure severe hardships; Officers and crew of the *Discovery* admitted that he had greater powers of endurance than any other man on board."[137] In such portrayals Scott's qualities as an explorer were based on both his physical strength and his ability to withstand hardship.

Even following his death, abstract qualities such as social status and self-control were often talked about through reference to the body and its organs. Cherry-Garrard assessed the ideal qualities of a polar explorer, explaining them in class-based terms; he claimed that "it was the sensitive men, the men with good nerves, with a background of education—'good blood'—who went farthest, pulled hardest, and stayed longest."[138] Again, what is interesting here is how these moral qualities were described through reference to specific organs.

Scott's dress was central to his heroic reputation as well. He was often depicted in naval uniform. Such images stressed his self-control and Britishness, conflating his individual body with the nation.[139] They also linked him to a longer tradition of British adventure in the polar regions.[140] In one widely circulated and often reproduced photograph, Scott was depicted in his dress uniform and appears a quintessential naval hero (fig. 5).[141] Class, authority, and heroism were very much embodied qualities. Here, though, the different ways that Scott dressed in the field and at home could bring to the surface divergent visions of what it meant to be an explorer.

Many pictures of Scott, Wilson, Evans, Bowers, and Oates portrayed the explorers in sledging gear. Such depictions emphasized the strains of the polar environment placed upon the explorers' body and show a more rugged vision of expeditionary heroism than images of Scott in uniform.[142] Scott's final message to the public drew on these ideas in a powerful way; he claimed that the expedition had "shown that Englishmen can endure hardships, help one another, and meet death with as great a fortitude as

FIG. 5. Captain R. F. Scott in naval uniform, photographed in 1905. The image was used as the frontispiece for the *Voyage of the "Discovery."* Wikimedia Commons.

ever in the past."[143] The "frenzied" response to Scott's message "exposes an age preoccupied with the endurance of hardship as a test of character."[144] Most importantly, a photograph of the men at the South Pole (similar to fig. 6) became an iconic symbol of the expedition's heroic sacrifice and featured prominently in the public and press response to Scott's death. The explorers' bodies played an important role in the emotional power of this image, particularly the "exhaustion and disappointment . . . engrained in the faces and bodies of the men who stare towards the operator-less camera, or at the vast expanse of snow beyond."[145] The photo acted as powerful visual proof of the hardships the men had endured; it is also an image quite different from images of Scott as the disciplined and controlled naval officer. Portraits of the explorers' weather-beaten and sunburned faces were a major component of the photographic record of the expedition.[146] Unlike Burton the explorers' darkened skin in the photograph caused few problems for their domestic reputations due to a growing view that "exposure to the elements was salutary for rebuilding manhood."[147]

We should not take such representations at face value. The various images of the ill-fated polar party only capture surfaces of their bodies, but the domestic viewer does not experience the smell of their unwashed skin, nor can they see the suppurating wound—full of "a lot of pus"—on the hand of Petty Officer Edgar Evans.[148] Death in the Polar regions is thus portrayed as a comparatively clean and hygienic way to die in which the white male body appeared to remain intact.[149] In order to turn the explorers into heroic icons, certain aspects of their bodies and experiences were ignored.

The different images of Scott in both naval uniform and sledging gear reflected competing ways of representing character, class, and control: one sought to emphasize Scott's disciplined naval body, while the other focused on a more private, personal vision of self-control illustrated by persistence in the face of physical suffering. Conflicting ideas about what it meant to be a heroic explorer continued to play out in discussions concerning the national memorial to Scott, eventually erected in London in 1915. This memorial was sculpted by Scott's widow, Kathleen Scott, and her private perspective on Scott's body shaped the way he was depicted. She drew on the widely circulated images of Scott in sledging gear when sculpting the statue, reflecting the broader popularity of the images. Her decision to portray Scott in this way was opposed by others involved in the memorial who felt he should be portrayed in naval uniform rather than "working clothes."[150] Even today the Scott memorial looks somewhat out of place, as the surrounding statues are all dressed in military uniforms (including the one of Franklin just discussed). Analyzing the different images of Scott brings to the surface underlying tensions in Edwardian visions of heroism.

Other classically "heroic" explorers saw their bodies interpreted in very different ways. The Burke and Wills expedition has many similar themes to Scott's later death. Robert O'Hara Burke and William John Wills led a party of nineteen across Australia in 1860–61, as part of an expedition organized by the Royal Society of Victoria. The return journey

FIG. 6. Scott, Wilson, Bowers, and Evans at the South Pole, 1912. Reproduced with permission of the Scott Polar Research Institute, University of Cambridge. Image ref: P2005/5/1345.

was plagued by disasters, and seven men, including Burke and Wills, died. Only one man—an Irishman named John King—crossed the continent and survived. The dead explorers' bodies received an incredible reception in Melbourne, and they were immediately portrayed as heroes. At the Royal Society in Melbourne, 120,000 people flocked to pay their respects to the explorers' corpses, and their hair and other relics were even put on display.[151] By the time a monument to the two men, depicting them in bush dress, was erected in the city in 1865, the British press felt no need to repeat the "heroic story of Burke and Wills" because it was so well known to readers.[152] Since then the story has been reinterpreted in various ways, reflecting changing attitudes toward heroism, nationalism, and exploration in Australia.[153]

But it is important to remember that, even in the 1860s, the expedition was not universally viewed as heroic. When news of the explorers' death reached Britain, some newspapers questioned the way the expedition had been organized and conducted. The *Dundee Courier* complained that "the whole affair from beginning to end has been a series of blunders and mismanagement." The newspaper suggested that the expedition's achievements, such as the crossing of the continent, had been purchased "at too high a price."[154] A five-man commission of enquiry chaired by British Army Officer Thomas Simson Pratt was established to investigate what had gone wrong. The final report questioned Burke's leadership and recordkeeping and deplored the "lamentable result" of the expedition, though it did praise his bravery and gallantry.[155] Nor was heroism evenly distributed within the party. The committee heard evidence that Charlie Gray, a working-class member of Burke and Wills's party, may have been a heavy drinker, and that this might have weakened his constitution, contributing to his death in 1861.[156]

Even David Livingstone, who went on to become an almost saintlike example of the ideal African explorer, was not always viewed this way. He undeniably was lauded for his self-reliance and self-control in the face of physical suffering, particularly following his death.[157] In displaying mental determination in the face of bodily hardship, he "exemplified the ideal of *character*" that was so prominent in the Victorian period.[158] But we should remember that he faced widespread criticism following his Zambesi expedition and his involvement in the ill-fated Universities Mission.[159] Even following his death, Livingstone's body, as Justin Livingstone has shown, was the site of different visions of heroism. Some writers emphasized his regional, national, or religious identity, while others focused on his scientific credentials. Many sought to combine these different qualities. Several writers drew attention to the loyalty he appeared to have inspired in seventy African men, women, and children who carried his body to the coast so that it could be returned to Britain. Yet only after the inspection of Livingstone's corpse was their account believed by some British commentators.[160] Livingstone's body played a central role in most of these narratives, which often portrayed him as heroic because he combined "manly vigour and spiritual virtue."[161] The multiple visions of heroism that emerged reflected broader social disagreements about what actions were most heroic and what the purpose of exploration was.

BEYOND HEROISM

We must move away from the language of heroism, as studying exploration through this lens can lead us to overlook explorers' more complex reputations. While they were alive, explorers were the subject of criticism as well as praise. Travelers knew this and represented their bodies in ways that they thought would advance their domestic reputations, but they also had to contend with the realities of performing a dangerous journey to a different location. In their public writings, they focused on their bodily attributes and used physical suffering to demonstrate mental strength. Both discipline and bodily breakdown could, accordingly, at different times and in different ways, be portrayed in positive terms. Dress was also a central issue to an explorer's reputation and could be used to emphasize their gender identity, cultural knowledge, or self-control and strength. Representations of explorers' bodies varied considerably depending on the time period in which they traveled, the location of the journey, and the identity of the individual in question. Explorers were not always successful in convincing domestic audiences that they behaved correctly, and strategies that gained support in one quarter attracted criticism in others. Representations of explorers' bodies were, therefore, far from monolithic: there were deep-seated disagreements about how a traveler should use and represent their body. Although explorers portrayed their experiences carefully, it often proved impossible to reconcile these broader disagreements. The same events could be portrayed in diametrically opposite terms.

Many explorers were seldom referred to as "heroes" in their own lifetimes. Individuals like Scott, Franklin, Burke and Wills, and Livingstone were widely lauded, but mostly so following their deaths. The real, fleshy, leaky body of the explorer in the field was transformed into an iconic body—and often memorialized through hard materials, such as bronze or stone. Memorials might thus be interpreted as an effort to imbue a given explorer's body with certain fixed meanings. But even dead explorers could find their bodies interpreted in quite different ways, both in their own lifetimes as well as by subsequent generations that reinterpreted their stories. Following the social and cultural changes of the 1960s, British culture took an increasingly critical attitude toward many of its former "imperial heroes." By the 1970s and 1980s, Victorian and Edwardian heroes became the objects of criticism rather than praise.[162] Such reinterpretations were possible because, in their own time, explorers were not understood in a monolithic way, especially in private.

While alive, explorers were celebrities.[163] Like modern celebrities they solicited both criticism and praise. In demonstrating the numerous ways that explorers' bodies were described and analyzed by contemporary commentators, this chapter suggests that even the singular body of an individual expedition leader could be read in multiple ways. The diverse reception of explorers' bodies was also a product of their fleshy materiality. While explorers could write about their travels in heroic terms, they could not always prevent incidents of illness and bodily breakdown that raised questions about their conduct and credibility. The importance of such incidents demonstrates the limitations of understanding explorers' texts as only literary sources. Explorers could polish and edit their accounts, but there were clearly limits to what was plausible or possible.

We should also move away from heroism for more fundamental reasons. In focusing on the individual body of (mostly white male) explorers, dominant visions of expeditionary heroism "excluded from the pantheon of heroes working men, professionals, women, and foreigners, including 'natives', and was shaped by and buttressed the prevailing rhetoric of imperialism, masculinity, and racial superiority."[164]

Understanding the relationship between bodies, heroes, and exploration has contemporary implications. Both critical geographers and institutions such as the RGS have begun to examine the organization's links with imperialism and colonialism.[165] Yet the RGS-IBG's South Kensington headquarters remains adorned, both inside and out, with images and statues of explorers' bodies.[166] The powerful role of the body within the mythology of exploration emphasizes the need for a more critical discussion about how contemporary geographical institutions explain and interpret statues, photographs, and paintings. Foregrounding the historical debates around explorers and their conduct might prove a way to challenge such static and one-dimensional images. It is the very fact that travel narratives are the site of "multiple meanings" that makes alternative readings possible.[167] If explorers were always heroes, there would be comparatively little to say about them and few possibilities for using them to tell different stories. Fortunately, this is not the case, and it is possible to use their writings to surface the narratives of a far broader range of experiences and contributions.

Even when we turn to this subject, ideas of heroism can be unhelpful. Explorers' accounts often draw on the myth of the "heroic indigene" in which individual subalterns are singled out for praise in ways that further obscure the collective work of exploration.[168] In his published account of the EAE, Speke did exactly this for one guide and translator: Sidi Mubarak Bombay.[169] Bombay is a remarkable individual. Of Yao heritage, he was

FIG. 7. Speke's "Faithfuls" as photographed at Cairo in 1863 by Louis Royer included in *The Journal of the Discovery of the Source of the Nile*. Courtesy of the New York Public Library. Image ref: 814735.

born in East Africa in around 1820. At age of twelve, he was captured and enslaved by Swahili-speaking coastal traders. He was then taken to a slave market at Kilwa on the East African Coast and sold to an Arab merchant, who took him to India. "I served this master for several years," Bombay reported, "till by his death I obtained my liberation."[170] Once free Bombay returned to Zanzibar and enlisted in the sultan's army. In 1856, while serving, he met Burton and Speke and joined the EAE. He had learned Hindustani in India, a skill that he used as a translator on the expedition, as Speke and Burton both spoke this language. Bombay also served on Speke and Grant's 1860 expedition and was photographed in Cairo in 1863 along with the other named "faithfuls," who remained with Speke throughout the journey. Bombay assumed a certain heroic reputation of his own and was awarded a silver medal by the RGS in 1863 before serving on other prominent expeditions, including those led by Henry Morton Stanley and Verney Lovett Cameron.

But the problem for contemporary historical geographers seeking to recover African contributions to European-led expeditions is that many descriptions of Bombay perpetuate negative stereotypes about Africans more broadly. Speke contended that Africans he encountered showed no interest in geography. He claimed that "Bombay, though, is a singular exception to this rule. . . . He has become a great geographer and delights in pointing out the different features on my map to his envying neighbours."[171] While, superficially, this quote seems positive, it perfectly exemplifies the challenges of using explorers' writings to foreground the contributions racialized people. If we focus only on more visible individuals such as Bombay, then we risk uncritically perpetuating their prejudices about Africans more broadly.[172] How, then, can we write about exploration while avoiding such pitfalls? As I now examine, focusing on a different aspect of the body can offer some solutions.

2

DEPENDENT BODIES

One of the most remarkable images in the archives of exploration comes from Ernest Shackleton's Imperial Trans-Antarctic Expedition (1914–17). The image, taken by the Australian photographer Frank Hurley, shows the expedition's scientists washing the floor of "the Ritz," their living quarters in the hold of the expedition's ill-fated ship, the *Endurance*. The photograph is powerful because it shows something quite fundamental about Shackleton's approach to leadership and management (fig. 8). On his expeditions no one was exempt from the manual labor—cleaning, cooking, and interpersonal care—that was vital to an expedition's success. It is perhaps this modern and egalitarian spirit that is responsible for the resurgence of interest in Shackleton in the twenty-first century in contrast to the imperial heroes discussed in the previous chapter.[1] An image of scientists might seem a strange place to start a discussion about the contributions of subaltern people to geographical exploration, but this photograph is useful because it reminds us that, under some circumstances, logistical work could be portrayed as a central aspect of exploration. As I show in this chapter, when comparable work was conducted by subaltern people on expeditions in different contexts, it was not nearly so visible, nor was it recorded in the same striking ways. The scientists scrubbing the floor reminds us to ask: Who did similar work on other expeditions? And why don't we celebrate their labor in the same way?

Over the past thirty years, historians and historical geographers have made increasing efforts to answer this question, studying individuals ignored within traditional histories of exploration.[2] They have focused on the contributions and agency of guides, intermediaries, and go-betweens who facilitated explorers' journeys and of working-class sailors to polar expeditions. This approach has allowed researchers to shift attention away from the heroic individual and toward the broader range of people who made expeditions possible.[3] Such projects have been described using

FIG. 8. Biweekly ablations of "The Ritz" on board the expedition's ship, the *Endurance*, 1915. Reproduced with permission of the Scott Polar Research Institute, University of Cambridge. Image ref: P66/18/42.

different terminology, as the "hidden histories" of exploration or through a focus on "indigenous intermediaries."[4] These terms are far from ideal, as many of the people who contributed to European exploration were "neither 'local' nor 'indigenous'" but were travelers themselves. The framework of "hidden history" risks "minimizing the difficulties of engaging with metropolitan archives" noted in the introduction to this book.[5] Some of the most exciting scholarship has examined the multiple forms of labor involved in explorers' journeys.[6] More popular literature and museum interpretations on the subject often approach this issue through the lens of heroism, by portraying guides, intermediaries, or working-class expedition members as the "alternative," "true," "unsung," or "super" heroes of exploration.[7] But what is the value of elevating a new group of heroes if, as I have argued, the framework of heroism is itself limiting?

In this chapter I focus on a broad range of individuals and expeditionary activities. I argue that, far from being simple, logistical work often involved skilled embodied labor and the cooperation of multiple bodies. I focus on these forms of work, such as transport, cooking, and care, to demonstrate the fundamentally "collective" nature of exploration.[8] In doing so I draw on the concept of "body work," a term that has a double meaning: it refers to "the work that individuals undertake on their own bodies, often through regimes of diet and exercise," and is also used to describe labor carried out on the bodies of others, such as interpersonal care.[9] I employ both meanings of the concept, showing that expedition members carried out work that depended on having strong, well-trained bodies, and that caring for the bodies of others was also central to the success of an expedition. I demonstrate that focusing on the body work of exploration provides a means to foreground a different set of experiences and contributions. These contributions are often missing from histories of exploration that focus only on activities directly related to knowledge production or specimen collection.[10] In short, concentrating on body work forces us to ask, "Who counts in the history of exploration?"

My strategy also provides a different approach through which to engage with explorers' writings and images, because they frequently mention expedition members' bodies—their strained muscles, sore feet, and caring labor—even when they omit important biographical details. Focusing on the body work of an expedition can, I argue, help us to write a different history of exploration that draws attention to a far broader array of contributions.

MUSCULAR LABOR

Everywhere explorers depended on the muscles of others to move their bodies, supplies, and equipment to and through the field. But the exact relations of dependance varied depending on where they traveled. Most European explorers in East Africa traveled in a preexisting caravan network powered by human muscles.[11] Porters (including some women) carried the explorers, along with their supplies and instruments. On the east coast of Africa, Nyamwezi porters from the central region of modern-day Tanzania formed the backbone of the increasingly professionalized labor force of the nineteenth-century caravan system. They therefore worked on most of the expeditions leaving from Zanzibar. Soldiers (*askari*) were also an important component of many parties, providing protection and sometimes carrying trade goods and equipment. Often these were Muslim men recruited in Zanzibar, but not always. On the East African Expedition (EAE) this role was performed by a group of Baluchi soldiers. Their title suggests that they were from Baluchistan, on the borders of present-day

Iran and Pakistan. Burton found that they included Afghans, Arabs, and men from across the Indian subcontinent.[12] Expeditions relied on both global and regional networks.

Some explorers hired huge labor forces. When Henry Morton Stanley's Emin Pasha Relief Expedition (1887–89) set out from Zanzibar it contained 611 porters, askari, and headmen. The group had "diverse origins and experience," containing men and adolescents from across Central and Equatorial Africa and included both Zanzibari and Sudanese men.[13] But even then the expedition's plans depended on recruiting more people en route. The disaster that befell this expedition illustrates the essential role carriers played in the progress and survival of an expeditionary party. The expedition traveled up the Congo from the west coast, reaching Yambuya (a point on the Aruwimi River, roughly equidistant between the east and west coasts of the continent). Stanley had brokered a deal with a local Arabo-African Ivory merchant who went by the nickname "Tippu Tip" to hire local carriers in the continent's interior to facilitate the expedition's onward march. But Tippu Tip was unable to provide enough men, leading to growing tensions with the expedition's leadership.[14]

Stanley divided the expedition, leaving a "rear column" behind, who were ordered to catch up when more porters could be recruited. He proceeded with a smaller advance column to relieve Emin Pasha at Lake Albert. Unable to move and suffering from increasingly poor leadership, the rear column was plagued by starvation, tropical diseases, and division. When he returned almost a year later, Stanley found that the rear column had only managed to move ninety miles. Worse still, 139 men (out of an original component of 270) were either dead, lost, or had deserted the expedition.[15] For William Bonny the only white officer of the rear column who survived and remained with the expedition, this experience forcefully illustrated the importance of African people to explorers' journeys. He wrote that, in the context of an expedition, "the black people are necessary to the white. . . . You could do nothing without him. You could not carry his load. If you took your grain and went to the next village without him you would get killed."[16]

In East Africa, too, explorers found recruiting and retaining a workforce a constant challenge. In one letter to the Royal Geographic Society (RGS) sent from East Africa, Speke complained, "I cannot move independently of the natives, and now the natives are not to be got for love or money [*sic*]."[17] Burton reported that he recruited four separate groups of porters for different legs of the EAE, and that two of his original party died of smallpox on the journey.[18] On his 1860 expedition with Speke, Grant recounted that 115 Nyamwezi porters set out carrying the explorers' supplies, but he struggled to retain a labor force due to political instability in East Africa at the time of his journey.[19] Explorers often sought to cast these desertions as a mark of indiscipline or disloyalty. In reality leaving a party when conditions became bad or near home was a normal part of the caravan system.[20] Expeditionary parties changed for other reasons. Many explorers recruited local porters for specific legs of the journey.[21] Guides, intermediaries, and porters bought enslaved people, including women and children, in the African interior, swelling the ranks of the expedition further.[22] The presence of such people illustrates the blurry boundaries of an expedition while raising questions about historical method, as recovering the biographies (and sometimes even the names) of many of them is impossible from the available sources—though explorers' roster lists offer some insights.[23]

Description of their labor demonstrate the important work they did. Porters carried an expedition's equipment. Stanley lists the variety of loads carried: "Cloth, beads, and wire, boat-fixings, tents, cooking utensils and dishes, medicine, powder, small shot, musket-balls, and metallic cartridges; instruments and small necessaries, such as soap, sugar, tea, coffee, Liebig's extract of meat, pemmican, candles, &c."[24] Each porter normally toted a load of about sixty pounds of trade goods. But, as the East African explorer Joseph Thomson noted, he might also have to haul his "personal property," such as sleeping mat, cooking pot, or gun, "which would commonly bring his actual load up to eighty pounds." A porter carried such loads "on his head or shoulders from two to eight hours a day." Large containers of beads were sometimes conveyed on poles lifted on the bare shoulders of two men.[25] The work of moving supplies was hard, collaborative, and skillful.

The fact that porters carried considerable quantitates of trade goods—fabric and beads—highlights the fact that explorers depended on people who were never formally part of the expedition. Few parties even tried to be self-sufficient when it came to supplies and equipment, and, as we have seen, many journeyed along preestablished routes of travel and trade. Along their journeys, therefore, they bought or were given food and other provisions by local people who had surpluses. I return to these relations in more detail in chapter 6, but it is worth noting here that such relationships throw into question the idea of "the expedition" as separate from the "local" people whom they encountered. Equally, acknowledging that explorers often relied on preexisting networks can throw into question the idea that many of their journeys were somehow remarkable or unique.[26]

While many explorers' journeys were far from unprecedented, they were certainly not easy, particularly for those who did most of the work. Carrying heavy loads required a strong, disciplined body, trained through many years of practice. It was not work that just anyone could undertake, a fact that is reflected in travelers' writings, though often in somewhat demeaning terms. Burton describes the bodies of the EAE Nyamwezi porters:

> The pagazi or porters are mostly lads, lank and light, with the lean and clean legs of leopards. Sometimes, however, a herculean form is found with the bullet-head, the broad bull-like neck, the deep wide chest, and the large strong extremities. . . . There is usually a sprinkling of greybeards, who might be expected . . . to be "leaning against the wall." Among these races, however, the older men, who have learned to husband their strength, fare better than their juniors, and the Africans, like the Arabs, object to a party which does not contain veterans in beard, age, and experience.[27]

There are clearly racist stereotypes at play in this description, but there are also elements of truth in the idea that many porters had trained their bodies over many years. Nyamwezi porters developed a unique and professionalized labor culture that celebrated physical strength and endurance.[28] Within this culture experience and the ability to "husband" this strength were resources that "could be drawn on in difficult times."[29] Participation in caravans became a "stamp of true manhood," and men who visited the coast would often change their names "to mark the significance of their visit."[30] As a result many were conditioned from an early age to become porters. Later European travelers to East Africa reported that Nyamwezi children practiced carrying loads. American explorer Mary French Sheldon noted that the Nyamwezi porters had "a muscular development in their shoulders in consequence of their habit of carrying loads upon them."[31] Grant, too, commented on the physical strength of

Nyamwezi porters, observing that the men were "very muscular, tall, fine men."[32] The labor that moved explorers through East Africa was far from simple and unskilled but was a product of years of body work.

Labor put severe strains on the porters' bodies. Burton reports that they often arrived at camp "with faces torn by the thorns projecting across the jungly path, with feet lacerated by stone and stub, and occasionally a leg lamed by stumbling into deep and narrow holes."[33] Speke noted several instances where he had to stop because the porters' feet were "sore" and "blistered."[34] While such brief passages give us a glimpse of the daily hardships of the porters, they also show how such suffering was seldom mentioned in explorers' accounts when it did not stall the progress of the expedition. When Speke returned to East Africa in 1860 with Grant, the porters' feet again suffered. Describing a revealing incident in which he hunted a giraffe, Grant noted that "our men benefited by the giraffe's thick skin, which they converted into excellent sandals," which were particularly welcome, as they were then traversing "thorny acacia country."[35] The fact that the porters had bare (and probably sore) feet on many other occasions is presented as unremarkable and hardly worthy of comment—nor do the explorers' private writings show much interest in the porters' experiences.[36]

Writings elsewhere in Grant's account help us understand why. Later Grant was forced to walk barefoot through a bog in Buganda. "Tears almost came to my eyes, the suffering from sharp roots was so severe," he wrote. Grant used this incident to draw attention to his own discomfort. He claimed that "the natives" had "feet as hard as leather" and therefore suffered less.[37] While the porters' feet may well have been tougher than Grant's, such comments draw on the racist idea common at the time (and still alive today) that Black bodies are less sensitive to pain.[38] Read more critically, though, it again gives us some clues as to the experiences of the East African porters. Walking hundreds of miles barefoot was clearly a painful undertaking that required specific physical and psychological qualities to which explorers' accounts only allude. Attention to this pain allows us to surface something of the experiences of the porters who carried the explorers and their supplies across East and Central Africa.

The explorers' own suffering may also have distracted them from the hardship experienced by those around them. Most explorers spent periods of their expeditions weakened by the effects of malaria, fever, and dysentery, and they were often too frail to walk unaided. As I examine further in chapter 4, the explorers rode donkeys and mules whenever they could. Most of these animals were soon killed by predators, sleeping sickness, and the tsetse fly.[39] Even when animals were available, though, the explorers were often too ill to ride. African porters (often but not exclusively Nyamwezi) therefore carried them for large portions of their journeys.[40] Focusing on the different ways that explorers wrote about such incidents in their published and personal writings demonstrates some of the difficulties of engaging with such texts. In private Burton described being carried on a stretcher in quite neutral language; he reported that he suffered from numb limbs and was "unable to walk or ride except in a hammock carried by Wanyamnezy [*sic*] porters."[41] In public, though, Burton portrayed such incidents differently. He claimed that the Nyamwezi porters who carried him on the return march were insensitive to his suffering and had to be bribed and coerced into moving.[42] In doing so he devalued the labor on which he relied.

Grant also describes being carried in his (ironically titled) book *A Walk across Africa* (1864). When leaving Karague he was too weak to walk and was conveyed by a group of men from the East African kingdom of Buganda. Grant complained that he found the stretcher uncomfortable, the stops too frequent, and that porters would not carry him feet first so that he could continue making observations.[43] The discomfort Burton and Grant experienced was perhaps because of their poor physical state while traveling, but it can also be seen as a rhetorical tool to manage their dependence on others. In focusing on their own experiences, the travelers drew attention away from the skilled embodied labor on which they depended, recasting their dependence as simply another source of suffering.

Few explorers were more dependent on the muscles of others than David Livingstone on his final expedition. Near death he was carried on both a stretcher and on the shoulders of African men. In his diary he expressed more sympathy for the men who moved him than most other British travelers: "Carrying me across one of the broad deep sedgy rivers is really a very difficult task."[44] Livingstone reported that lifting him involved the collaborative labor of numerous men; he had to be passed from one pair of "willing shoulders" to another.[45] Each man could only carry him fifty yards before getting out of breath: "No wonder!" he commented.[46] In contrast to most writers, Livingstone names most of those

FIG. 9. Captain Grant leaving Karague, from Speke's *Journal of the Discovery of the Source of the Nile*. Speke did not include an image of himself being carried. Courtesy of the New York Public Library. Image ref: 1258411.

who carried him and shows a greater appreciation of the labor involved. This serves to illustrate how hard and poorly acknowledged this work was on other expeditions. But even Livingstone's more sympathetic telling gives us very little information about the experiences and motivations of the men who carried out this labor.

When Isabella Bird traveled through China in 1895–96, she relied heavily on the muscular labor of Chinese men. As others have noted, white women travelers were often able to assume relations of domination over racialized men because their racial identity was more important than their gender identity in the colonial context. During her fifteen months in China Bird directly employed 134 people.[47] Of these we only know the names of a handful of men, such as Bird's personal servant, referred to as "Be-dien," who accompanied her for most of the trip. Most others are unnamed in her account, but, again, a critical focus on their labor allows us to see something of their contributions and experiences.

For much of her journey, Bird traveled up the Yangtze by boat while human labor pulled her vessel upstream. Bird refers to the men who towed her boat as "trackers."[48] Bird reported that, at the start of the journey, her

FIG. 10. David Livingstone, suffering from fever, carried through a river on the shoulders of one of his men. Lithograph. Courtesy of the Wellcome Collection. Image ref: 561398i.

boat had sixteen trackers on board—but the boat's captain recruited more en route to help pull the boat up particular rapids. These men who were so central to the success of her journey are generally anonymized in her account, but she did take the time to photograph them and to describe their meals, noting that they ate only "inferior rice mixed with cabbage or other vegetables fried in oil, with a bit of fish or pork occasionally added. . . . On three such meals," she continues, "the poor fellows haul with all their strength for twelve hours daily never shirking their work."[49]

Bird describes the "inhuman work" of the trackers when hauling the boat through rapids, noting that they pull the boat from the shore attached to a rope, often in time with a drum. The men "swing the arms and body forward, stooping so low to their work that their hands nearly touch the ground, and at a distance they look like quadrupeds." Bird observed the effects of this work on their bodies, claiming that "severe strains and hernia are common, produced by the tremendous efforts in dragging."[50] Moving the boat up one "very severe rapid . . . took us four hours of desperate dragging," she wrote. Bird's use of the word "us" is revealing, as she did not personally do any pulling but was "sitting shivering for that time on a big boulder."[51] Bird's descriptions of the trackers highlight her dependence on their hard and disciplined muscular labor. She rhetorically manages this dependence by both emphasizing her own suffering and by describing the trackers bodies using animalizing language. As elsewhere Bird's brief comments allude to a tremendous amount of labor but the identities of those who performed it are largely obscured.

Similar patterns of dependence are visible on Bird's overland journeys. Three separate groups of Chinese men carried Bird (and her supplies) 1,200 miles across China in an open chair.[52] Bird described the bamboo chair, suspended on 14-foot poles, as "light" and "comfortable" but reported that it carried "forty pounds of luggage," including her camera (see fig. 12). Bird's longest description of the men who carried her depicts the first group of chair-bearers recruited: "Two were opium smokers, and the third was a vegetarian, who abstained from opium, tobacco, and *samshu* [an alcoholic beverage made from rice]." Bird said that the bearers carried her for distances of "thirty miles on some days."[53] Occasional references to the chair-bearers scattered throughout *The Yangtze Valley and Beyond* (1899) highlight the degree to which Bird depended on their skilled, embodied labor during the journey and shed some light on what this entailed:

> Chair travelling is, I think, the easiest method of locomotion by land. My one objection to it is the constant shifting of the short bamboo carrying pole on which the long poles hang, from one shoulder of each bearer to the other. It has to be done simultaneously, involves a stoppage, occurs every hundred yards and under, and always gives the impression that the shoulder which is relieved is in unbearable pain. Chair-bearing is a trade by itself, and bearers have to be brought up to it. It is essential to keep step absolutely, and to be harmonious in all movements.[54]

FIG. 11. Authors' trackers at dinner, photograph by Isabella Bird Bishop, Fukien, 1895. This image was published in her book *The Yangtze Valley and Beyond*. Reproduced with permission of the RGS-IBG. Image ref: S0011249.

In referring to chair traveling as "easy," Bird draws attention to her own experiences of the journey. But the brief mention of the bearers' sore shoulders suggests that that the men's experiences differed significantly from hers. The importance of synchronized movements also illustrates the skilled nature of this work. Even from such one-sided passages, we see that moving her chair involved muscular labor and coordination—again, the product of many years of training.

In these descriptions most of the body work of exploration was done by subaltern men. In Antarctica, though, there was no indigenous population—and British men did far more of the hard muscular labor than in other contexts. As elsewhere human muscle remained an essential component of most expedition's transport plans.[55] Transporting large loads again required physical strength, skill, and a high degree of coordination. Effective sledding necessitated the men to move their bodies together, with each member pulling "rhythmically" at the right time.[56] Robert F. Scott reported that, done effectively, the individual explorers' bodies were transformed "into a machine—a machine that must keep moving

FIG. 12. Mrs. Bishop's (Isabella Bird's) arrival with baggage coolies at a country inn, Chekiang, 1895, photograph by Bird. Reproduced with permission of the RGS-IBG. Image ref: S0010964.

with that regular swinging step . . . and can do so without straining its parts."[57] If one member was struggling, or pulled at the wrong time, then sledding became hard, inefficient, and uncomfortable.[58]

When the men of the British Antarctic Expedition ascended the Beardmore Glacier in 1911 on their way to the South Pole, a team led by Edward Evans, the expedition's second in command began to struggle. Scott swapped places with Evans and found that "the team was not swinging with me as my own team swings," suggesting in his diary that "bad stepping and want of swing" might have been the source of the problem.[59] While the issue turned out to be with the sledge rather than the men, the quote highlights how the explorers had to learn and discipline their bodies to "swing" together in order to move the sledge forward. The embodied labor of sledding was collective, and the explorers depended on each other in new ways. These descriptions are suggestive of the skilled muscular labor performed by porters and chair bearers, as discussed. But what the Antarctic example also shows is how differently such labor was valued when it was carried out by white men narrating their own experiences in another environmental context.

FIG. 13. Sledge party on the march, British Antarctic Expedition. Reproduced with permission of the Scott Polar Research Institute, University of Cambridge. Image ref: P2005/5/1267.

It is important not to overstate the British valorization of "man-hauling" sledges.[60] As I will examine, Robert F. Scott tried to use animals or machines to ease some of this labor: on his first expeditions he used a group of much-suffering huskies, and on his second he used dogs, ponies, and motor tractors with somewhat more success.[61] Nevertheless, it is clear that the labor and suffering of the white male body in Antarctica could be an important part of what made exploration heroic.[62] In part this was because polar exploration was a direct response to fears about physical fitness and racial stamina, and a way of proving the continuing vitality of the British body.[63] Following Scott's death Sir Clements Markham, former president of the RGS, praised the muscular labor of man-hauling, claiming that, "with men dragging their own laden sledges," Scott's journey was "by far the finest sledge journey that has been achieved in either of the polar regions."[64] Similarly, the British magazine the *Bookman* argued that Scott and his companions had "more dearly won" the prize of the South Pole than their Norwegian counterparts (who had beaten them to the Pole by getting dogs to pull their sleds) because on their journey they were "straining nerve and muscle to the utmost gain their rightful laurels."[65]

Antarctic expeditions are thus useful for exploring the devaluation of muscular labor in other contexts. When such work was performed by white men, it was transformed from menial logistical work into an essential (and even heroic) aspect of expeditionary labor. That said, there were still inequalities in how such work was represented, particularly along the lines of class. Some who reported on the death of Scott's party suggested that the largest man, Petty Officer Edgar Evans, had been the first to die on the return from the Pole because of the inability of working-class sailors to discipline their minds and bodies as effectively as officers.[66] As I will show in chapter 3, officers' published narratives often described working-class sailors in patronizing terms. In all contexts exploration depended on the muscular, embodied labor of numerous people. The exact relations of dependence were heavily influenced by local environmental and social conditions, but how such labor was valued varied considerably, shaped by the identity of the people performing it. Similar dynamics emerge in descriptions of caring labor.

CARING, COOKING, AND CLEANING

Relations of care and dependence are another important form of body work. To remain healthy travelers needed food, clothing, shelter, and interpersonal care, particularly when they fell ill. Caring labor is often "dirty work" involving "close and intimate work on the bodies of others."[67] These forms of labor are often gendered, with women generally doing a disproportionate share.[68] On expeditions class, gender, and racial inequalities shaped how these activities were divided. Drawing attention to cooking, caring, and cleaning highlights how the scientific and adventurous elements of an expedition were made possible by the work of others. Again, though, the visibility of and importance of such tasks was shaped by questions of identity.

On British expeditions in Africa and Asia, how much we know about the individuals who cared for explorers varies considerably. When traveling through Persia, Bird accompanied a "military-geographical" expedition led by Major Sawyer, a British intelligence officer.[69] For nine months of the journey, Bird hired Mirza Yusuf, "an educated young Brahmin, for some years in the British post-office service in the gulf, and lately a teacher."[70] She also hired a Persian cook called Hassan. Yusuf and Hassan cared for Bird's body. She wrote that her servants "all pitch tents, and Hassan cooks, washes, packs the cooking table equipments [*sic*] and saddles my horse. Mirza Yusuf interprets, waits on me, packs the tent furnishings, rides with me, and is always within hearing of my whistle . . . and though well educated had the good sense not to regard any work as 'menial.'"[71] Bird's reference to her whistle illustrates the almost constant support she expected from those around her. The list of activities the men performed is also extensive. At the same time, Bird managed her dependence by presenting the work on which she relied as "menial."

Likewise, Bird reported that Be-dien, who performed a similar role in her Chinese travels, "was always at hand to help me over slippery and difficult places, showed great pluck, never grumbled, arranged and packed up my things, interpreted carefully, improved fairly in English, always contrived to get hot water and food for me, and on the whole made a tolerable travelling servant."[72] In this sentence Bird both acknowledges Be-dien's work but also portrays this important caring labor as the work of a "travelling servant" rather than an explorer or expedition member—and only a "tolerable" one at that! Care is thus both central to Bird's journey but treated as a private and nonscientific form of work.

The work of cooking and caring was often highly skilled. Such talents are occasionally alluded to in Bird's writings. In Korea Bird was accompanied by "a Chinese servant," to whom she referred to by the name "Wong." She condescendingly described him as "a big cheery fellow, with inexhaustible good-nature and contentment, never a cloud of annoyance on his face, always making the best of everything, ready to help every one [*sic*]." Elsewhere, though, we get more sense of his skilled work on the journey. When traveling by boat, Bird reported that "Wong used to hunt along the river banks for wild onions and carrots, after the stock of cultivated roots was exhausted, and he made paste of flour and water, rolled it with a bamboo on the top of a box, cut it into biscuits with the lid of a tin, and baked them in the frying pan. Rice fritters too he made morning, noon, and night."[73] The caring labor of exploration was, therefore, far from unskilled or simple and, in fact, required ingenuity, culinary skills, and botanical knowledge.

In other instances we do not know the names or identities of those who cared for Bird. For instance, in *Among the Tibetans* (1894), Bird gave an account of her typical marching routine:

> I sent a coolie carrying the small tent and lunch basket to await me half-way. . . . On reaching the shelter tent, I halted for two hours, or till the caravan had got a good start after passing me. At the end of the march I usually found the tent pitched on irrigated ground, near a hamlet, the headman of which provided milk, fuel, fodder and other necessaries at fixed prices. "Afternoon tea" was speedily prepared, and dinner, consisting of roast meat and boiled rice was ready two hours later.[74]

Bird included a very similar description of an ordinary day's march in her published account of her Persian travels, suggesting that this was the routine that she sought to impose on her expeditions.[75] Bird's use of the derogatory term "coolie" and the passive voice obscures the identity of those who did this work and the complexity of these tasks carried out. Bird clearly had many of her bodily needs met by others, but she left us with little information on those who did such labor.

Bird's writings about similar activities in other contexts shows how such "domestic" work could be portrayed very differently. In the American West, Bird deliberately sought out opportunities to cook and clean. She viewed this labor in idealized terms: "If my back gets well enough, I seriously think that a servant's place would be the best thing. Manual labor a rough life and freedom from conventionalities and added to novelties would be a good thing" she claimed before the trip. Clearly, this is a unrealistic picture of what life as a servant would be like, as Bird began to realize. During about her time in the Rocky Mountains, where she stayed in a hut with a group of white men, Bird emphasized that she took on duties elsewhere performed by working-class women: washing dishes and clothes, cleaning, and baking. "I now even understand the management of an American cooking stove," she commented in a letter to her sister Henrietta.[76] Travel thus saw Bird take on forms of work that she considered unacceptable for middle- and upper-class women at home. She included descriptions of this labor in her published account, allowing her to perform a rough-and-tumble version of English femininity.[77]

Explorers might also use descriptions of care to portray themselves as benevolent. Bird often provided medical attention to local people when she traveled in Asia and the Middle East, treating minor skin infections and wounds.[78] Describing such incidents allowed Bird to depict her perceived racial and cultural superiority and the benevolent motivations for her journey. In *The Yangtze Valley and Beyond*, Bird reported that she tended to a severely ill Chinese man in her party bathing his head and walking to allow him to ride on her mule, emphasizing how her compassion contrasted with the "brutality" of the Chinese men she was traveling with who, she claims, were willing to allow the man to die. She thus renders caring for others as a display of morality and kindness—which serves to illustrate how subjective the valuation of this labor could be. But Bird's compassion had limits. Sympathy for the sick porter did not stop her photographing the man, nor from including this image in her book on the journey.[79] Caring work was not always devalued, but explorers wrote about it in ways that upheld racial and cultural hierarchies.

Masculine explorers might outsource such caring duties. When traveling in the Amazon in 1848, the naturalist Henry Walter Bates became concerned when his mixed-race guide José fell ill. Feeling unable to care for him, Bates went to a local settlement to "procure aid." There he hired an elderly woman to take care of him. He reported that "she made poultices of the pulp of wild fruit, administered cooling draughts made from herbs which grew near the house, and in fact acted the part of the nurse admirably."[80] The incident not only illustrates how guides and other intermediary figures were often in need of as much care as expedition leaders, but it also shows that explorers unfamiliar with the regions through which they were traveling sometimes lacked the skills to provide such care. Instead they drew on the botanical and medical expertise of local people.

Similar imperatives were at play on African expeditions. On the EAE Burton and Speke employed two Goanese servants, Valentine Rodrigues and Gaetano Andrade, to work as valets and cooks. In public Burton complained about the inconsistent standard of their catering, remarking that their diet "varies from such common doings as the holcus-scone, the tasteless bean-broth, and the leathery goat-steak, to fixings of delicate venison, fatted capon, and young Guinea-fowl or partridge, with 'bread sauce,' composed of bruised rice and milk."[81] Despite Burton's grumbles

Rodrigues and Andrade continued to cook and care for him and Speke while suffering from fever themselves. Rodrigues also acted as a tailor and was, Burton claimed, "as clever at sewing a coat as at cooking a curry" and made clothes for both Burton and Speke.[82] The two Goanese men also cared for the explorers when they were ill. Burton noted that Andrade "had a curious kind of tenderness when acting as a nurse."[83]

Interpersonal care when an explorer was sick was a key form of expeditionary labor. Sidi Mubarak Bombay also played an important role in caring for Burton and Speke. On one march Burton was suffering severely from fever but had to walk because of a shortage of transport animals. "Towards the end of that long march," he reported, "I saw with pleasure the kindly face of Seedy [*sic*] Bombay who was returning to me in hot haste, leading an ass, and carrying a few scones and hard-boiled eggs." Bombay also drew on his own experiences of illness to help Speke. On the return march, Speke had a fever known as the "little irons"; Bombay (who had already had the same illness) "raised his master's [Speke's] right arm, placed him in a sitting position, as lying down was impossible, and directed him to hold the left ear behind the head, thus relieving the excruciating and torturing twinges by lifting the lung from the liver. The next spasm was less severe."[84] Bombay's assistance in this situation demonstrates his detailed understanding of the fever and its effects. It also brings to the surface the weakness of Speke's body, which appears almost infantilized by his total prostration. To reassert control explorers often emphasized their mental determination in their narratives. For instance, Speke later wrote that he "thought of nothing but the march" despite his physical collapse on his 1860–63 expedition.[85] But it is hard to read such descriptions of total bodily dependence without suspecting that, in the field if not on the page, those who cared for explorers had a significant power over the bodies (and lives) of expedition leaders.

These relationships of tenderness and care were not reciprocal. Speke reported that in July 1858 Gaetano, suffering from malaria, "slipped away behind the caravan, and hid himself in the jungle to suffer the pangs of fever in solitude." Speke referred to this action in animalizing and disparaging terms, suggesting that the man had "like a dog, slinked off to die."[86] Gaetano was eventually found by a search party and recovered. But this near-death experience illustrates the different way that illness was valued in travel narratives. If the roles had been reversed—as they so frequently were—Speke would have narrated his own suffering and departure from the party very differently.

Other African explorers were cared for in similar ways to Burton and Speke. On his final expedition, Livingstone's body required particular care. He had lost many of his teeth and had a bowel problem, limiting the kinds of food that he could chew and digest. His cook, Halima, went to great lengths to provide him with suitable food, carefully grinding "fine flour, such as she only could prepare" so that he could eat hot food. Yet Livingstone apparently complained that Halima "never can tell the difference between tea and coffee."[87] It is not hard to imagine that the food eaten on expeditions was often far from perfect, given the cooking facilities available. But such complaints also work to devalue the contributions of those who did much of the work of exploration.

Again one of the challenges of recovering the caring labor involved in an expedition is the shifting of labor forces over a single journey. When Speke returned to East Africa in 1860, two Cape Mounted Riflemen, privates Vandermerwe and April, were employed as cooks and valets until they fell sick, and Speke sent them back due to ill health. Rahan and Frij, Swahili-speaking men, originally hired as interpreters, took over most of the cooking duties for the rest of the expedition.[88] Both men's backgrounds stretched across East and Central Africa. Like other Waungwana they had been captured and enslaved as children and transported to Zanzibar, where they were brought up as Muslims. Though formally enslaved they carried out wage labor and played a prominent role in the East African caravan system.[89] As Muslims they ensured that all food prepared on the expedition was halal.[90] The two men were assisted by a "cook boy" called M'kate, about whom we know little from a biographical perspective. Grant described him as "a very tall good-looking lad, ever obliging and good humoured." Grant complained about the team's cooking, but he also related a revealing incident. He reported that M'kate "one day left a cooking pot twelve miles behind. He was admonished by Frij and took the matter so much to heart that he travelled back for it alone that same day and returned during the night, having recovered the old pot, which was certainly not worth the journey. . . . He was not asked to go back, and had walked by himself 36 miles in a strange country."[91]

Grant diminishes M'kate's contribution by suggesting that the pot was not worth the effort, but, read more fairly, the incident highlights how the expedition's cooks took a pride in their work and cared for their equipment even in the difficult circumstances of the expedition. It is not hard to imagine that if such a journey had been undertaken by a white explorer to recover a scientific instrument, it would have been regaled into altogether more heroic terms.

It was not just the bodies of expedition leaders that required shelter, nourishment, and care. On many expeditions porters ate and slept in groups, thereby benefiting from economies of scale. Joseph Thomson was impressed by the organization of the porters on arriving at a camp. He noted that tasks were divided, so that one man would light a fire to cook dinner, another would forage for food, while still others would assemble a shelter. "The rapidity with which a whole camp was run up was really marvelous," he commented.[92] Another strategy adopted by porters, guides, and headmen to lighten their loads was to purchase enslaved people or to hire servants of their own to relieve them of some of their work.

A focus on care also brings to the surface the presence of women in expeditionary parties. Women, both free and enslaved, played a key role in the nineteenth-century caravan system. They had varying degrees of power: many were the partners of the Nyamwezi and Waungwana porters, sometimes as a result of temporary "caravan marriages."[93] Porters also purchased enslaved women and children in the African interior.[94] They played a variety of roles on the caravan, carrying loads, provisions, and supplying domestic services and sexual relationships. Their contributions are harder to establish from the available sources due to the reticence of European explorers to talk about their relationships with African women or their complicity in slavery.[95] We do find occasional references to their presence and their work (and I examine this issue further in chapter 7).[96] Grant notes that "a few" women traveled with the Nyamwezi porters, "generally carrying a child each on their backs, a small stool and et ceteras on their heads, and inveterately smoking during the march. They would prepare some savoury dish of herbs for their men on getting into camp, where they lived in bell-shaped erections made with boughs of trees."[97] Focusing on relations of care thus draws attention to the presence of women in expeditionary parties, as well as to the limitations of explorers' writings as a source of information about them. These complex caring relationships illustrate the limits of thinking about status and labor in expeditionary parties in binary terms. Some subaltern people had much more power than others, and we often know very little about the identities of those with less power.

In Antarctica we know far more about the names and identities of those that did labor such as cleaning and cooking. There were significant differences in how explorers approached such activities. Scott's expeditions followed in a longer tradition of hierarchical Royal Naval exploration in the polar regions, in which sailors performed a disproportionate share of embodied labor.[98] When in base camp or on board ship, much of this work was carried out by working-class men.[99] Writing in his diary from the *Terra Nova* expedition (1910–13), Scott argued that this division of labor was important to their scientific productivity: "I think it is a good thing in these matters the officers need not wait on themselves; it gives long unbroken days of scientific work and must, therefore, be an economy of brain in the long run."[100] The cooks and "domestics" on Scott's expedition are also visible in photographs of the expedition (fig. 14). Scott saw their labor as valuable (if not equally so). But this division did contribute to the exclusion of working-class people from more heroic acts of scientific discovery.[101]

Shackleton, as mentioned, took a more egalitarian approach to such activities. On his *Nimrod* expedition (1907–9), he ensured that all "had a certain amount of work for the common weal." In the expedition's hut at Cape Royds, the explorers took turns acting as "nightwatchman" and "messman." The nightwatchman took care of the hut, ponies, and stove overnight. The duties of the messman were "more onerous" and involved laying the table, boiling water, washing up dishes, and sweeping up the hut.[102] Of course this should not be taken as a sign that these expeditions were anything except hierarchical; even where such work was more evenly divided, leaders became far more famous for their activities than others. In both examples, though, what is different in Antarctica is the importance and value given to such work in published accounts. Here interpersonal care and domestic labor was the work of explorers, not "servants."

Polar expedition also saw men carrying out new forms of labor to create a homely environment.[103] These activities brought the explorers' bodies into close proximity as well. Cutting hair was a particularly collaborative undertaking. Raymond Priestley, who traveled to Antarctica with both Scott and Shackleton, noted that "every man was his neighbour's

FIG. 14. Clissold making pies on the British Antarctic Expedition, 1911. Reproduced with permission of the Scott Polar Research Institute, University of Cambridge. Image ref: P2005/5/477.

barber."[104] Again the photographic record of polar exploration includes a remarkable array of images of explorers engaged in interpersonal care (see fig. 15). These new configurations meant that officers also began to recognize the complexities of some types of labor that they had previously taken for granted. On both of Scott's expeditions, officers washed their own clothes.[105] "I shall always have a high respect for laundry-work

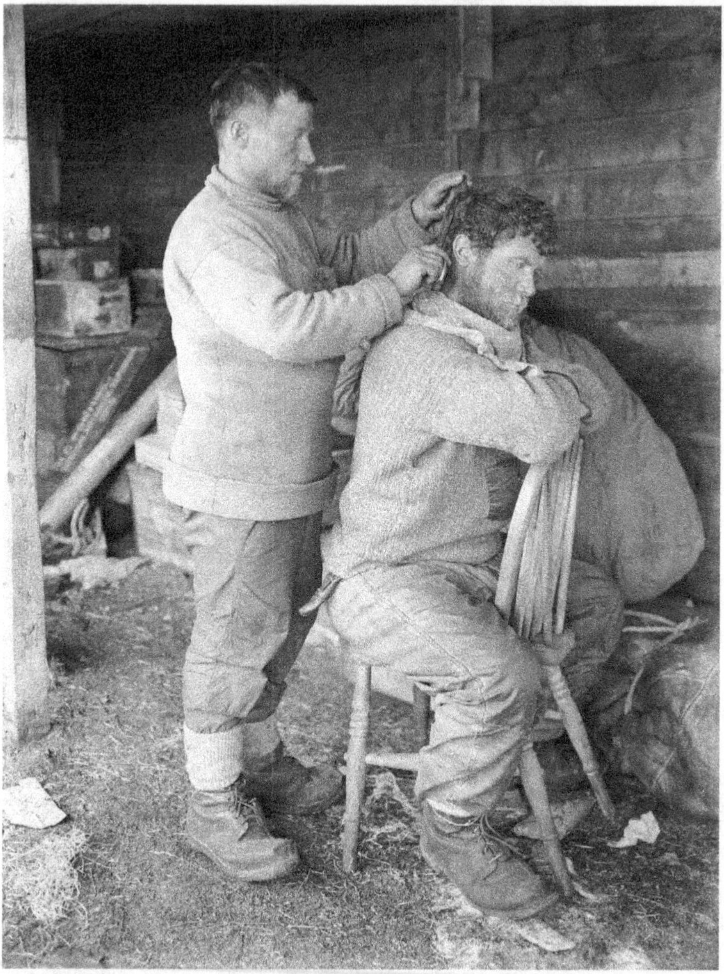

FIG. 15. Hair cutting on the British Antarctic Expedition. Reproduced with permission of the Scott Polar Research Institute, University of Cambridge. Ref: p2005/5/833.

in future," Scott wrote after the experience.[106] Carrying out such work could lead officers to develop a newfound appreciation of its difficulties. The fact that naval officers partook in work that was, in other contexts, divided along class, racial, or gender lines made these tasks more visible in their published narratives. Priestley even included images of such work in his book on the expedition.[107] The visibility of such activities in explorers' published accounts shows that domestic work, when carried out by certain people in certain spaces, could be rendered a central part of expeditionary activity. In Antarctica "there was nothing unmanly about cooking and serving food to his fellows, boasting about his handiness with a darning needle, or leading a sing-song over a smoke."[108]

The collaborative work of cooking is particularly visible in explorers' accounts of sledding. While traveling they rotated cooking duties, normally taking weekly turns to heat food and water for the party's three daily meals.[109] This was the first time some upper-class expedition members had cooked for themselves. Anxious about the prospect of catering for his tentmates, Apsley Cherry-Garrard even took lessons from the chef of his stately home before joining the *Terra Nova* expedition.[110] Cooking in such situations certainly was a work of skill, and in their accounts of their expeditions explorers commented on the quality of the food prepared by different individuals, suggesting that it made a marked difference to how full they felt.[111] Attention to the difficulties of cooking in a polar tent could even lead to these activities being portrayed as one of the heroic difficulties of polar travel. Scott included a vivid description of polar catering in *The Voyage of the* Discovery (1905) emphasizing the challenges of handling metal pans with bare hands and rationing out food supplies.[112] He also noted that the men had to move their bodies carefully in the confined space: "Elbows and knees have to be managed with caution to avoid disaster to the cooker."[113] Similar descriptions appear in other explorers' published accounts.[114] As such the labor of cooking is presented in a level of detail absent in other contexts and is rendered an essential activity (see fig. 16).

Antarctic travel also relied on numerous forms of interpersonal care.[115] Such were the strains of polar travel that the breakdown of one expedition member could threaten the safety of the whole party. Antarctic sledgers thus devoted considerable attention to looking after each other's bodies. Wilson's diary from the *Terra Nova* expedition contains numerous passages where he comments on issues such as frostbites, blisters, snow blindness, and infected cuts.[116] Treating such ailments meant that interpersonal care was an important part of the men's daily routine.[117] For instance, explorers often warmed frostbitten parts of their companions' flesh using the heat from their own body. Shackleton reported how important this was when the cold, starving men were confined to their tents during a blizzard: "Every now and then one of our party's feet go,

FIG. 16. Cooking food in a tent on the British Antarctic Expedition, 1911. This is a staged shot, as it would not have been possible to take with the tent fully erected. Reproduced with permission of the Scott Polar Research Institute, University of Cambridge. Image ref: p2005/5/1695.

and the un-fortunate beggar has to take his leg out of the sleeping-bag and have his frozen foot nursed into life again by placing it inside the shirt, against the skin of his almost equally unfortunate neighbour."[118] Sharing bodily warmth was thus an important survival technique.

Polar explorers sometimes even slept together in three-person sleeping bags, as these were warmer than one-person bags. The middle position was the most comfortable, because that man would be warmed by the bodies of the other two men. But this close proximity was not entirely pleasant. Wilson noted that the smell produced by men wearing "clothes which they have been sweating freely every day for week, can be more easily imagined than described."[119] Scott reported that sleeping in such bags blurred the boundaries between the men's bodies in profound ways. When it was cold, one man might start to shiver. The effect of one man shivering was "contagious" and soon passed through the other men in the bag. At the same time, Scott viewed this loss of control as a "necessary part of the programme" because of its "warming effect," and on very cold spring nights the men often shivered for "an hour or two" before sleeping.[120] The men's bodies were pushed into interdependent relations of close proximity.

Just surviving while sledding relied on the explorers cooperating and looking after each other, coordinating their movements and showing close attention to the health of those around them. Descriptions of such work were included in explorers' published narratives. These instances show how interpersonal care could be rendered a central and heroic aspect of an expeditionary journey. Polar travel put such a strain on the men's bodies that cooperation was a question of survival. In Antarctica labor that, in other contexts, was dismissed as the work of "traveling servants" was rendered as an essential aspect of exploration. Again this different approach seems to have been largely motivated by questions of authorship, location, and identity.

CONCLUSIONS

Concentrating on the labor involved in caring for and nourishing the expedition's bodies allows us to see exploration in different terms. Producing credible geographical knowledge required explorers to be in a reasonable physical state. Keeping explorers nourished and healthy under the difficult conditions of an expedition involved demanding work and complex relationships of dependence. Putting this work at the center of analysis highlights how scientific research and geographical discovery were only possible because of the caring labor of others. In many explorers' published and private accounts, such work is often recast in ways so as to either obscure the identity of the people who did it or to downplay its value.[121] Though, as identified earlier, there are sometimes important differences in how explorers wrote in public and private. The approach emphasizes the importance of embodied labor in the history of geographical exploration; it also challenges the idea that fieldwork was conducted by solitary, individual bodies. People interacted and became interdependent in complex ways. Studying all those involved in each expedition allows us to see it in different terms and to move away from focusing on heroic individuals. Multiple bodies and multiple kinds of works are brought into view.

Care and logistics were far from simple or unskilled. In fact such work often required cultural, medical, and botanical knowledge, as well as experience and discipline. In recovering these contributions, the task is often developing the correct interpretive approach so as not to perpetuate the class and racial prejudices that permeate explorers' writings.[122] Such interpretation is important because the value attached to such labor varied considerably. Explorers often devalued embodied work when it was carried out by subaltern men and women. When similar work was carried out by white men, it was treated as an essential aspect of exploration (and perhaps even as heroic). That said, hierarchies of labor still existed in Antarctica, with working-class men performing a greater share of such work, particularly on Scott's expeditions. Gender differences were also important. White women travelers might represent care in different ways, using it to demonstrate their femininity and the humanitarian motivations for their travels.

It is possible to tell different stories about the history of geography through a critical reading of the public and private records of exploration. Although these sources are problematic, it is possible to "surface" subaltern bodies even when the names and biographies of such individuals are unknown.[123] Important in this process is providing a close, contextualized reading of explorers' accounts. A focus on embodied labor can provide a means to foreground subaltern contributions and experiences. But concentrating on contributions can draw attention away from issues of agency and motivation.

Reading explorers' accounts, and the disparaging way they often wrote about those whom they depended on, a reader is often left wondering: What motivated African and Asian expedition members to join and stay with an explorer? One incident pushed these issues to the fore more than any other: the death of Livingstone. After Livingstone died in Chitambo (in modern-day Zambia) in 1873, his body was carried over one thousand miles to the coast by a diverse group of around seventy African men and women who had formed his expeditionary party. Most well known among them are James Chuma, Abdullah Susi, and Jacob Wainwright, who achieved a significant degree of fame and recognition in Britain (see fig. 17). But the party also contained women, including Halima, Livingstone's cook (discussed earlier). In both historical and contemporary accounts, they are often referred to as Livingstone's "servants," but, as elsewhere, they carried out forms of work that would have been referred to as "exploration" if undertaken by white men. As with incidents described

FIG. 17. James Chuma and Abdullah Susi with relics of the Livingstone Expedition, photographed at Newstead Abbey, Nottingham, 1874. Courtesy of the Wellcome collection. Image ref: 561462i.

in this chapter, the story of how Livingstone's body was preserved and transported to the coast, highlights the caring and muscular labor, skill, and strength that many African expedition members exemplified. The party went to great lengths to preserve Livingstone's body after his death, salting and drying his corpse.

But the incident also brings up questions around agency and motivation. Many who have read or heard the party's account have been left with the same question: Why did these men and women go to such great efforts to transport his body to the coast? A variety of explanations have been suggested. The answers they have come up with reflect different perspectives on both African agency and wider colonial power relations.[124] Many Victorian accounts, which emphasized the group's loyalty and devotion to Livingstone, depicted Africans as "faithful servants" or "native attendants" in passages laden with colonial tropes.[125] Their labor was used in ways that drew attention back to Livingstone's apparent leadership qualities rather than the agency of those who performed the work of preserving and carrying his body. More recent writers have suggested alternative explanations, such as a fear of being blamed for Livingstone's death or the hope of financial reward.[126] These explanations have their own problems, as the journey was itself highly risky, and prospects of financial reward slim, and Livingstone's journals (which some in the party could read and which showed how ill Livingstone was) would have been enough to remove any suggestion of foul play. We are therefore left with more questions than answers. But what this remarkable journey does show is that there were often forces holding an expeditionary party together beyond the physical presence of a living leader. It is to these issues of agency, discipline, and motivations—and the difficulties of researching them through explorers' writings—that I now turn.

3

DISCIPLINED BODIES

The American arctic explorer Robert E. Peary likened the ideal expedition to "the physical structure of a tough, hardy man. . . . Following this analogy, one intelligent white man would represent the head." Other white men "selected solely for their courage, determination, physical strength, and devotion to the leader would represent the arms." He then claimed that "the natives" were "the body and the legs."[1] This metaphor provided a means to simultaneously acknowledge and manage the contributions of others to his expeditions. To him subaltern work was important, but those who did it had to be under total control of the expedition leader. Similar views can be found in the writings of African explorers. Henry Morton Stanley described himself as "the vanguard, the reporter, the thinker, the leader of the Expedition."[2] Such ideas on expeditionary discipline were an extreme expression of a broader consensus within exploration cultures: the success of a journey depended on the ability of the leader, generally a white man, to turn the multiple parts into a coordinated whole. Subordinate expedition members were either portrayed as docile "instruments"—tools that could be used and discarded—or as unruly obstacles to the progress of the expedition.[3] In this chapter I argue that writing about discipline allowed explorers to rhetorically manage their dependence on others.

From a different perspective, this subject can tell us about the agency and contributions of the broader range of people involved in expeditions. Explorers frequently found themselves with limited control over either their own bodies or those they nominally commanded. Leaders often relied on the self-control of subordinate members and on preexisting and culturally hybrid forms of discipline. This chapter builds on the previous chapter by moving beyond discussions of multiple contributions to address the significant role that subaltern agency played in Victorian and Edwardian exploration. But I also show the problems with taking this argument too far, as many expedition leaders relied on violence—sometimes to the point of murder—against the bodies of racialized people. Discipline thus brings to the surface broader issues of power and identity.

Efforts to regulate and control bodies are a long-standing feature of Western society. Premodern ideas were shaped by religious notions about the need to resist temptations of the flesh.[4] The nineteenth century saw profound changes in European attitudes toward discipline. Bodies were "manipulated, shaped [and] trained" by these regimes of power—a process that both developed and contained their capacities—within the military, schools, factories, asylums, and hospitals.[5] The perfectly disciplined body was the "analyzable and manipulable" and "docile body."[6] Discipline was thus a quality that blurred the boundaries between mind, body, and behavior. Different things were also expected of leaders and the led.

For an explorer to be in command of an expedition, they had to be in control of themselves. Advice to explorers emphasized the importance of self-control and restraint. Richard Burton suggested that "study and discipline" were essential qualities in any "first-rate traveller."[7] Francis Galton advised explorers that an "even temper" was crucial when dealing with subordinates, and that "great allowance should be made to the reluctant co-operation of servants."[8] These views were quite common. Before his departure to East Africa in 1860, Speke claimed James Grant was an ideal companion because of "his conciliatory manner with coloured men" and "for his general good temper & patience."[9] Here it is the patience of the expedition leader, rather than the led, that is vital.

One technique explorers used to discipline their minds and bodies was exercise. Joseph Thomson, another African explorer, claimed that "the only way to resist successfully the enervating effects of humid tropical climates is by constant exertion, and by manfully fighting the baleful influence." Failing to keep mentally and physically busy, he argued, would result in a man degenerating "into an idiot or baby," leaving them unable to lead a party. "Hard constant work is the great preserver," he summarized.[10] Similar ideas were shared by Arctic explorers, who saw exercise as a key means to preserve health on board a ship.[11]

Others saw strength and self-control as qualities only possessed by certain kinds of bodies. Sir Clements Markham, the president of the Royal Geographical Society (RGS) between 1893 and 1905, viewed naval officers as having attributes that rendered them suitable for exploration. During the selection of the leader of the British National Antarctic Expedition (BNAE), he argued forcefully that "all previous good work in the polar regions has been done by young officers in the regular line."[12] He claimed that older officers were "physically unfit" for polar leadership, listing a long list of previous expeditions where the appointment of men over forty had led to disaster.[13] Markham's views were clearly not universally held—otherwise he would not have had to lay them out so explicitly. But they do illustrate that some saw expeditionary leadership as a matter of body as well as mind.

While the ideal explorer remained in control of their own body, real travelers struggled to stay healthy. This discrepancy is particularly visible in the case of East African explorers.[14] Although they adopted a variety of measures to maintain their personal health, protecting themselves from the sun, drinking medicinal brandy, as well as taking tablets laced with quinine and the opiate laudanum, they spent much of their expeditions physically feeble and mentally anguished.[15] Burton's private correspondence from the East African Expedition (EAE) often emphasized his weakness. In one letter to the RGS written from the Lake Regions, Burton complained of both an inability to walk and of the "terrible wear & tear of mind."[16] Elsewhere he described one fever lasting twenty days in which he experienced "a queer conviction of divided identity, never ceasing to be two person that generally thwarted and opposed each other" as well as vivid and disturbing hallucinations.[17] These experiences were far from unusual. Stanley reported the disturbing mental and physical effects of fevers, commenting that "the brain becomes crowded with strange fancies, which sometimes assume most hideous shapes."[18] Thomson wrote about the effects of malaria in very similar terms.[19] African explorers spent much of their time "out of their minds" with illness, fatigue, and anger.[20]

Nor was such mental and physical anguish a quirk confined to African exploration. On polar expeditions scurvy, darkness, hunger, and exhaustion changed the relationship between the mind and body. Scurvy was a common complaint on British expeditions in this period. The disease caused a range of disturbing physical and mental symptoms, including hemorrhages, tooth loss, the reopening of old wounds, depression, sensitivity to loud noises, and irritability. Robert F. Scott suffered with scurvy on the southern journey of the BNAE. His account chronicles how the men's minds gradually escaped their control.

Scott reported that fantasies of food—and the feasts they would have when they returned to the ship—consumed more and more of their thoughts and conversations, and that they were "powerless to prevent it." Their empty stomachs appeared to be shaping their behaviors, too. At one stage they were unable to save their lunch for the appropriate time. Scott wrote that they ate it early, as "the temptation became too great; though we knew it was wrong, our famished condition swept us away, and in five minutes not a remnant remained intact."[21] When they eventually reached a depot with more plentiful food supplies, they found themselves unable to contain themselves, eating pot after pot of "hoosh," biscuits, jam, and cocoa—so much food that both Scott and Edward Wilson, the expedition's junior medic, were physically sick.[22] The ideal expedition might well have a leader who exemplified physical strength and mental self-control. But, in the field, explorers found it very hard to maintain such standards. So how did explorers maintain some semblance of authority when they were so often so out of control themselves?

TROPES OF DISCIPLINE AND DISORDER

One strategy explorers used was to present themselves as more physically and mentally resilient than those they commanded. In their published accounts, explorers often represented those they led as undisciplined, at least without their influence, implying that whatever qualities subordinates might have, it was the leader who drove the expedition forward. What is remarkable is how similar these writings could be, even when explorers were commanding very different groups of people.

Throughout their travel narratives explorers often portrayed those they led as either childlike or animalistic and used this indiscipline to display their own heroic qualities of management and determination. Speke, for instance, portrayed African men as like children. In *Journal of the Discovery of the Source of the Nile* (1863), he wrote in disparaging terms about the lack of self-control of the Waungwana men who formed the nucleus of his expedition, declaring that "laziness is inherent in these men. . . . Although extremely powerful, they will not work unless compelled to do so." Drawing on broader racial and climatic stereotypes, Speke claimed that the average Mwungwana was "a creature of impulse—a grown child" and asserted that "economy, care or forethought never enters his head." Speke also suggested that this lack of discipline extended to their sexual and moral behaviors, stating that their "greatest delight is in the fair sex . . . next comes beer, song and a dance."[23] He included numerous accounts of drunken squabbles among the African members of his expedition, playing into broader racialized tropes of African indiscipline. But Speke claimed that such men can be "trained to work" through a combination of "great forbearance, occasionally tinctured with a little fatherly severity."[24]

Such descriptions are fairly typical of African explorers, who often used the trope of the childlike African to emphasize their own qualities as leaders. Thomson drew on similar language in his own account of East African exploration. He reported that a group of porters were pretending to be unwell to try and secure a day's rest from the march. In response Thomson apparently administered a dose of the laxative castor oil to each man with "parental sternness" and threatened to administer another if the men were still unwell the next morning. This treatment proved so effective that the party, he claimed, "marched out of camp merrily" the next morning.[25]

The explorer-as-parent trope was drawn on by white women as well. Mary Kingsley frequently referred to the African men she led using the patronizing term "boys," describing them in ways that conform to racist stereotypes about African people. In her account of the ascent of Mount Cameroon, she depicts the African men with her as hapless and cowardly, suggesting that they required "intellectual support in the matter of procuring firewood," and that she had to stay awake to prevent them from burning themselves in the resultant fire. Kingsley portrayed her authority as emanating not from violence but from the paternal loyalty she inspired in those around her. Summarizing her summiting of Mount Cameroon, she claimed: "The only point I congratulate myself on is having got my men up so high, and back again, undamaged; but, as they said, I was a Father and a Mother to them, and a very stern though kind set of parents I have been." Like male explorers Kingsley characterizes African men as useful followers, but incapable of leadership, making the statement that efforts by African servants to manage things generally "produce such awful results."[26]

On polar expeditions naval sailors were also portrayed as simpleminded, though in slightly different ways. In *The Voyage of the* Discovery (1905), Scott reported that the main problem with many of the BNAE's working-class sailors was their inability to admit that they were worn out. In the expedition's second sledging season, he and a party of men journeyed to reach the polar plateau. Scott intended to send some of the sailors back and to proceed with only the strongest men. None of the sailors wanted to return, he claimed, viewing it as a sign of weakness and failure. "What children these men are! and yet what splendid children! They won't give up till they break down, and then they consider their collapse disgraceful," Scott exclaimed in his published account.[27] He portrayed the sailors as unable to fully understand the physical limits of their own bodies, suggesting that they need his paternal guidance. These descriptions are subtly different from the tropes of the childlike African just discussed: the sailors are too keen to work rather than too unwilling.

But Scott still saw the sailors as lacking his own foresight and self-control. He criticized their "wasteful use of food, clothes, and equipment." After describing various attempts to reduce their consumption, he concluded that it might be impossible "to alter their happy-go-lucky-nature."[28] While less racially charged than the writings of African explorers, his conception of exploration, as a fundamentally hierarchical undertaking, is similar.

White women did not write about commanding white men in the same way. As we have seen, such women often wrote about adopting positions of paternal dominance over African men, but they were reluctant to write about the issue in the same way when they were leading European men. In *A Lady's Life in the Rocky Mountains* (1879), Isabella Bird described an unsuccessful expedition—made by Mr. and Mrs. Chalmers, a settler couple, and Bird—toward Estes Park in 1873. Bird emphasized that she initially took a subservient position, allowing the man to guide the party. She described how Mr. Chalmers "who had started confident, bumptious, blatant, was ever becoming more bewildered."[29] Soon they were lost and without water, leading to falls and injuries. Eventually, when Chalmers confessed he was lost, Bird stated that she had "much experience in travelling and would take control of the party."[30] Bird perhaps felt confident to assume control of the working-class couple because of her upper-middle-class background. Even so, what is striking is how long Bird claims to have waited before she began to lead, despite Chalmers's incompetence and the danger it posed. The incident shows that Bird treated situations where she asserted her control over white men with extreme caution in her public writings.

Dealing with disobedient subordinates was an established—and even expected—feature of exploration narratives. Leaders often used similar language. They almost always portrayed subordinates in certain terms: childlike, shortsighted, cunning, or naively faithful. These prejudices took different forms in different contexts, but the idea that subordinates' bodies needed discipline was remarkably consistent. Of course, enforcing discipline in the difficult conditions of an expedition could be challenging. Yet we also see that explorers' writings also reflected domestic cultural expectations, literary tropes, and explorers' own prejudices as much as the reality of exploration.

SUBALTERN SELF-DISCIPLINE

While explorers might portray those they commanded as inherently undisciplined, there are good reasons to doubt many of their claims. A careful reading of explorers' published and private writings demonstrates that they often depended on the self-control of those they commanded. This was not accidental, as leaders often went to great lengths to select disciplined expedition members. Through a focus on such attributes, we begin to see something of the skills and agency of the various people on whom explorers relied.

Explorers often sought to recruit a small, disciplined group of men who would form the core of their expeditionary party. Speke, for instance, declared "that men accustomed to discipline" should be enlisted "to give confidence to the rest of the men."[31] On the numerous British expeditions that left Zanzibar for the lake regions of Africa, Waungwana played this role. Such men, as noted, had generally been enslaved as children in the East African interior before being moved to the coast and brought up as Muslims. Many, but by no means all, had then been formally liberated from slavery, while some also served in the Zanzibari military. Their experience of traveling and trading in the East African interior as well as their familiarity with several languages made them important "cultural brokers" within the caravan system. Waungwana often worked as *askari* (soldiers), headmen, and valets, giving them a more senior (and often better paid) status compared to the Nyamwezi porters.

These men were often disciplined and motivated, excited by the personal and financial opportunities that travel into the East African interior offered.[32] Speke was impressed by the Waungwana employed on his 1860–63 expedition, at least in private. In one unpublished letter, he noted that they do "all the work and do it as an enlightened and disciplined people."[33] Several of these men went on to work for other explorers, including Stanley and Verney Lovett Cameron. Stanley was also impressed by the disciplined conduct of these men, but depicted it in racially offensive terms, writing of their "useful docility" compared to the "negro savagery" of men from the interior.[34] We should clearly reject the prejudice in such statements, but they illustrate that Waungwana men had qualities that explorers admired.

Effective relationships relied on leaders respecting Muslim men's religious practices and social status. Islam was an important aspect of Waungwana identity. In public explorers were often dismissive of their knowledge about and belief in Islam.[35] But the Waungwana went to significant lengths to follow Muslim dietary rules, turning down incorrectly prepared meat, even when hungry.[36] Explorers often adapted their hunting methods so that the throats of animals could be cut, rendering the meat halal.[37] When the Waungwana found that some meat they were hoping to consume had spoiled, Speke noted that the others accused the cook Frij of having "unclean hands," because they quite reasonably thought that "unless the man's hands are pure who cuts the throat of an animal, its flesh will not last half the ordinary time."[38] Expedition leaders might have liked to see themselves as in total charge of a party, but much of the day-to-day conduct of the men they commanded was shaped by ideas about conduct and discipline over which they had little control.

Expeditions also included large numbers of porters. As discussed, Nyamwezi men did much of the work of moving a caravan into the East African interior. As skilled and often sought-after laborers, it was generally porters who had the final say whether they took part in an expedition, with their motivations shaped by shifting political and economic conditions in the region.[39] In *How I Found Livingstone* (1873), Stanley reported that when he tried to recruit porters at Bagamoyo on the coast, several came to his camp to see how heavy the loads would be. When they proved to be weightier than they considered acceptable, they refused to join his party, seeking work elsewhere.[40] Stanley was forced to repack his loads into lighter parcels. The porters, not Stanley, decided much of the day-to-day arrangements for the journey.

In the 1860s and 1870s, before European colonies were established in East Africa, porters were often able to leave an expedition or strike for higher pay and better conditions. But while porters had autonomy, they were far from undisciplined. They had preexisting structures that made it possible to control large bodies of men on difficult overland journeys. The caravan's *mtongi* "oversaw the allocation of loads." This task was a "work of skill," as "each individual . . . must choose or, at any rate, consent, to his burden."[41] The "*mganga* or traditional doctor and diviner, another important caravan official, acted as advisor and provided ritual protection against the dangers of the road."[42]

Kirangozis played a particularly essential roles in expeditionary discipline. They were elected leaders who guided the march, navigating it along the correct route, marking a trail which the other porters would follow.[43] The kirangozi's dress played a prominent role in instilling the status of the expedition on surrounding villages and among its members. They would lead the march, dressed in "fantastic costumes" and "wonderful headdress," carrying a "blood-red" flag as a sign of the Zanzibari Sultan's authority.[44]

Explorers also described the important role of the kirangozi in maintaining discipline, at the same time reinforcing wider tropes about the difficulty of controlling African porters. According to Thomson they were "looked up to by the men as their advisers on all questions affecting the caravan." He also noted that the kirangozis "considered all orders the leader may have to make," such as the route, the length of march, quantity of rations. "Whatever their decision," he continued, "it is implicitly followed by the entire caravan."[45] Burton wrote that "the only way of breaking the perverse and headstrong heard into a semblance of discipline, is to support the kirangozi at all conjunctures, and to make him, if possible, dole out the daily rations and portion the occasional presents of meat."[46] Thomson advised future explorers of the vital importance of getting "an influence over these men. It is only in this way they can hope to get along pleasantly."[47] Stanley also relied on the party's kirangozi to give orders to the porters.[48] Such quotes demonstrate that the kirangozi took the lead on routine disciplinary matters and show the dependence of expedition leaders on their good will. Caravan figures therefore played a

FIG. 18. Caravan figures as depicted in Burton's *The Lake Regions of Central Africa*. Note the presence of a mother and child. Courtesy of New York public Library. Image ref: 1248356.

leading role in directing the expedition's movement through East Africa and in maintaining order (see fig. 18).

Relying on preexisting caravan networks of Nyamwezi porters could take explorers most of the way into the African interior. But it was often hard to recruit such men away from the main trade routes or in regions where travel was particularly dangerous. Many explorers employed African children who, they hoped, would be easier to command. Often these children had been enslaved by Swahili or Arab merchants. Some were then purchased and freed by explorers on their journeys; others were liberated by British vessels working to suppress the slave trade off the East African coast. After being freed many went on to work for the explorers who had "liberated" them. Often their linguistic skills meant that they carried out important work as translators. Some children went on to become highly skilled travelers as adults.

James Chuma was released from slavery at the age of eleven by David Livingstone in 1861 on the Zambesi Expedition. After traveling to India with Livingstone, Chuma was educated in Bombay. He was then employed on several further expeditions in East Africa, including Livingstone's last expedition, where he carried Livingstone's body to the coast following his death. Notably, Chuma traveled with Thomson, on an RGS-sponsored expedition that trekked into the interior of the continent from the East African coast (1878–80). Thomson's descriptions give us some sense of Chuma's skills as an adult. Chuma was "acquainted with English, and about a dozen native dialects." He was also an impressive orator and used rousing speeches to drive the porters forward. Chuma was also full of "anecdote, and fun, and jollity," but he "preserved such an authority over them [the porters] that no one presumed to disobey his orders." Because the experienced, culturally knowledgeable, and linguistically adept Chuma was the "headman" of the party, the expedition had few problems with indiscipline.[49]

So great were Chuma's skills that he even ended up instructing Thomson, who confessed that part of the role of a "headman" was the "education of the white traveller" in the skills of exploration. "The headman may practically do what he likes with his pupil," he noted. Based on Thomson's descriptions it seems hard to overstate the important role that Chuma played in the daily control of the party. Perhaps aware that his audience might ask questions about who was in charge, Thomson made efforts to reassert control. In terms loaded with racial prejudice, he claimed that Chuma was habitually dishonest and lavish in his expenditure, and that Chuma was an "invaluable helper" when managed by a European.[50] With this rhetorical flourish, Thomson reduces his role from coleader of the expedition to a comparatively junior functionary. Read more fairly Chuma's contributions raise questions about who was leading whom.

Chuma had many years of experience to draw on when working with Thomson. But even comparatively young and inexperienced guides could contest the authority of an expedition leader. Bird's writings about her Japanese travels in 1878 provide a compelling example of how complicated power relations on an expedition could be, even when explorers employed young men as guides. On her journey through Japan, Bird hired an eighteen-year-old named Ito Tsurukichi as her guide. In her book on the journey, *Unbeaten Tracks in Japan* (1880), she emphasized her discomfort with a position of management over her guide: "I have never been able to manage anybody in my life, and will surely have no control over this clever, cunning Japanese youth."[51] As the comment about his "cunning" nature shows, Bird's approach to him was heavily shaped by her bigotry. Her discussion of leadership is best viewed as a literary trope inserted to show her feminine discomfort with authority. Bird was a relatively experienced traveler and, as discussed, had assumed command in demanding situations.

In many passages Bird claimed that, under her guidance, Tsurukichi had acted a faithful and docile subordinate.[52] In private she stressed her dominance over him, suggesting that it was "wonderful" that she felt safe to travel with "only a boy of 18 as interpreter."[53] By refusing to recognize Tsurukichi as an adult, Bird emphasized her authority over him. She described how, in public, too, she succeeded in getting him to stop wearing a hat that she thought "obnoxious."[54] More broadly, "the image of Ito that emerges . . . is of a capable, dependable 'boy,' a servant over whom Bird is able, from an early point, to gain control."[55] Bird's writings about Tsurukichi stress her ability to discipline his body, turning him into a useful instrument of exploration. Yet even in her descriptions of his service, we see her reliance on his self-discipline. She lauds the fact that he avoids alcohol, takes "great pains to be accurate," and sends any spare money back to his mother.

Bird's account was not one-dimensional. Other passages in it raise questions about who was in charge. Elsewhere in her narrative, Bird suggested that the relationship between the two of them was far from a clear hierarchy. "I am trying to manage him, because he meant to manage me," she reported.[56] She also had concerns about the fact that Tsurukichi apparently beat his horse "unmercifully" and called him a "bully" and a "coward" for doing so. "I am now making Ito ride in front of me," she continued, "to make sure that he does not beat or otherwise misuse his beast." Initially her description of this incident shows how she used the order of march to instill her authority and to ensure that animals were controlled in humane ways. Tsurukichi's response, as recorded by Bird, illustrates his agency and self-control. She wrote that, when they next stopped, he "brought out his notebook, and quietly asked me the meanings of the words 'bully' and 'coward.'" He used Bird's attempts to punish him as an opportunity to improve his language skills, a move that inverted the power relationship, causing Bird "aggravation."[57] Because her authority was based on negotiation and persuasion, it was open to challenge through questioning and misunderstanding.

The conflict over authority was particularly visible around the issue of manners. She reported that Tsurukichi "delights in retailing stories of the bad manners of Englishmen" and "is very anxious about my good behaviour, and as I am equally anxious to be courteous everywhere in Japanese fashion, and not to violate the general rules of Japanese etiquette, I take his suggestions as to what I ought to do and avoid in very good part, and my bows are growing more profound every day!"[58] These comments suggest that, while in Japan, Tsurukichi attempted to regulate Bird's manners and conduct to comply with local codes of etiquette. She implied that she did actually alter her habits in order to avoid causing offense, suggesting that the power relationship in the field was not always the clear master-servant binary that Bird elsewhere presented it as. Previous scholarship has indicated that these contests can be read as reflection of broader Western anxieties about the emerging power of Japan.[59] Studied alongside other expeditions, it is clear that contested power relationships were a common and productive feature of Victorian travel and exploration.

Like explorers in East Africa, Bird relied on more formalized, preexisting disciplinary structures when she traveled in China. Their important and productive role is particularly visible in her account of her trip up the river Yangtze by boat. The river was a long-established trade route; even the dangerous Three Gorges had been open to "regular shipping" for nearly one thousand years by the time she traveled there. The waterway was also governed by an established system of customs and regulations that made her movement possible.[60] In much of Bird's book, the role that such structures played in her ability to travel is not obvious.[61] Her account of the ascent of the Hsin-tan rapids demonstrates how the disciplined and coordinated bodies of the boat's crew played a key role in her movement up the river. The rapids were a dangerous stretch of water that the boat had to cross to proceed inland. She described how local pilots helped her vessel to navigate:

> A curious functionary came on board my boat, a well-dressed man, carrying a white flag; on which was written, "Powers of the waters, give a lucky star for the journey." He stood well forward, waving

> this flag regularly during the ascent to propitiate the river deities, and the cook threw rice on the billows with the same object. The pilot was a quiet, well-dressed man, giving orders by signals which were promptly obeyed. Indeed, the strict discipline to which these wild boatmen submit in perilous places is remarkable.[62]

The boat's ascent of the rapids was possible because of ordered and coordinated labor that required many years of training. Bird was not even on the vessel at this time and generally walked on the shore alongside the river, as she was "nervous" of rapids after an accident in Korea.[63] The movement of the boat up the river was carried out by an orderly crew over which Bird had almost no control at that crucial point.

Not all expeditions were run on such formal lines. When he traveled on the river Amazon, the naturalist Henry Walter Bates was a passenger on board small cargo-carrying vessels that traversed the river. Such boats often had diverse crews, including settlers, Indigenous people, and people of mixed heritage. Bates often struggled to understand how such vessels were run. "The Indians [*sic*] forward had things pretty much their own way. No system of watches was followed; when any one was so disposed he lay down on the deck and went to sleep; but a feeling of good fellowship seemed always to exist amongst them."[64] On another vessel he was struck by the fact that the Indigenous crewmen treated the ship's officers "with the most amusing familiarity, yet I never sailed in a better managed vessel."[65] Bates was probably drawing on colonial stereotypes about the laid-back nature of Indigenous people in these descriptions, and the apparent lack of formal discipline may have reflected his own ignorance rather than the actual way the boats was run. Nevertheless it is plausible that the crews had genuinely found ways to run a boat efficiently and without recourse to harsh disciplinary structures.

For many of these explorers, the main challenge they faced was recruiting people to join their party and preventing them from leaving. Maritime polar expeditions, in contrast, often brought large crews of sailors to the polar regions. Once there it was difficult (and often impossible) for unwell or disgruntled expedition members to leave. The leaders and organizers of such expeditions thus had somewhat different worries from other explorers. They were concerned less with desertion than with ill health, poor morale, and mutiny. The Admiralty had long been aware of this and thought that "undisciplined crews" were one of the main "dangers" of polar exploration.[66]

Most of the time, British polar expeditions were crewed by volunteers. Organizers of expeditions therefore went to great lengths to select disciplined, healthy expedition members. On the British Arctic Expedition (BAE, 1875–76), the leader, George Nares, ensured that all crew were between twenty-five and thirty-two years old, avoiding younger men with little service record so that an individual's "medical and moral" history could be ascertained. All men were medically examined and were generally between five feet five inches and five feet eight inches tall "in order to prevent obtaining heavy men, or men of too little strength."[67] One officer noted that "the slightest defect, such as bad teeth or old wound, was sufficient for refusing the service of otherwise strong men."[68] Other explorers employed similar measures. Before the departure of HMS *Erebus* and HMS *Terror* from Greenhithe in 1845, Sir John Franklin had several men discharged on account of their poor health, presumably discovered through a medical examination.[69] Time and effort went into to selecting disciplined crews for polar expeditions.

Similar methods were employed on the BNAE. All crew were volunteers, warned on application that they would be "selected for their physique, conduct, ability, and recommendation, but a clean medical history is essential."[70] Medical examinations were conducted both on board the ships from which volunteers were enlisted and by the Admiralty.[71] Sailors with bad teeth or other health problems were rejected.[72] Unlike on the BAE, a number of sailors were large, having impressively (perhaps even ideally) muscular bodies. Albert Armitage, the expedition's second-in-command, later claimed that the "chest expansion of some of the crew was very fine indeed," and that their physical proportions would have "called forth admiration from the Sandowists."[73] Armitage's reference here is to followers of Eugen Sandow (see fig. 19), "the father of modern bodybuilding and the first man to make a comfortable living by displaying his muscular physique."[74] The image of Sandow gives us some sense of how muscular some of the sailors must have been. The physical prowess of the *Discovery*'s crew was a product of working-class body work and self-mastery.

Different standards applied to the expedition's officers and scientific staff. They were examined separately.[75] Some were treated more leniently

FIG. 19. Eugen Sandow photographed between 1900 and 1909. Courtesy of the Wellcome Collection. Image ref: 2046759i.

in respect to their physical strength than the sailors. Wilson was accepted, despite suffering with lung problems, due to Scott's intervention.[76] The expedition's third lieutenant, Ernest Shackleton, managed to avoid examination altogether—which was fortunate for his career as a polar explorer, as he had a heart condition. Officers and scientists could secure places while having less impressive bodies. That said, this was not the case when the expedition reached Antarctica, and Scott later sent Shackleton back to England on the basis that officers should be fitter than the men they commanded, as they may be called on to undergo "hardship" and "exposure . . . at any moment."[77] Before the expedition sailors required stronger bodies than officers, but, once in Antarctica, Scott viewed the issue in quite different terms.

On paper naval expeditions were governed by strict regulations. In the first half of the nineteenth century, the navy's Articles of War, significantly reworked in 1749, governed discipline on board naval vessels, allowing a range of violent punishments, including flogging and execution. In 1860 these articles were updated by the Naval Discipline Act, which set out a broad array of punishable offenses, including drunkenness, sodomy, and uncleanness, and regulated public worship.[78] The punishments laid out differed for officers and men and generally reduced the use of flogging.[79] Sailors were more likely to be punished through imprisonment or detention, while officers were more likely to be dismissed.[80] By the early twentieth century, the act had been amended several times to restrict (but not totally ban) the use of violent punishments.[81]

While formally governed by such strict regulations, discipline on naval expeditions was also a product of the self-disciplined conduct of the crew. As we have seen, naval explorers were volunteers selected on account of their good service records, so it is hardly surprising that officers were often impressed with their conduct. Franklin wrote to a friend, noting that the officers of his 1845 expedition were "a fine set of young men, active zealous, and devoted to the Service. Equally good are the crew—and many say that no ship could go to sea better appointed than we are."[82] Likewise, Commander James Fitzjames described the *Erebus*'s crew as "fine hearty fellows" and noted their generally good conduct when some men were let ashore in Stromness in Orkney.[83] Albert Markham was also impressed by the sailors of the BAE, observing that they were "always to be depended on in a crisis" and, while sledding worked with "a zealous desire to do their best."[84] Sailors were often highly self-motivated.

On the BNAE, the crew and officers were mostly drawn from the navy, but were not officially bound military discipline, as the *Discovery* was not a naval ship, and the crew were employed through civilian contracts. Scott did not think civilian regulations strict enough for a polar service, so the explorers "lived exactly as though the ship and all on board had been under the Naval Discipline Act."[85] Reflecting naval customs, officers and men ate, slept, and socialized separately on board the *Discovery* (although they ate largely the same food). By continuing to observe far stricter regulations, even though there was no legal requirement for them to do so, the crew contributed to discipline on the expedition.[86]

Because it had less naval involvement, Scott's second expedition was even more reliant on the self-discipline of working-class expedition members. The composition of this expedition was more varied, with fewer sailors and a greater number of civilian scientists. As a result daily life in the hut was characterized as much by a culture of joking and care as much as by naval regulations and self-mastery.[87] That said, certain forms of naval hierarchy were transposed to the expedition's hut. The interior was divided between officers and men, with a partition between the two sections. "I am quite sure it is to the satisfaction of both," Scott commented in his diary.[88] The sailors' views on the matter are harder to recover.

The lack of formal discipline on Scott's second expedition could cause problems, although explorers' accounts are circumspect in their discussions of such issues. Apsley Cherry-Garrard reported that until the last year (following Scott's death) the explorers relied on volunteers for many tasks "that should have been organised as routine." The result, he claimed, was "that the willing horses were overworked."[89] From other accounts it seems that the sailors who conducted themselves in a "conservative naval way" carried out much of this labor. The expedition's second-in-command, Edward Evans, complained that some mostly upper-class civilian scientists "had never come under any sort of naval or military discipline" and lacked the sailors' self-control.[90] The presence of such descriptions in Evans's account was no doubt shaped by his own naval background and stereotypes of heroic naval sailors.[91] But Scott also hints at his general reliance on their self-control, expressing his surprise when a job was not done with "seamanlike care."[92] While a criticism this quote suggests that, the rest of the time, he expected particularly high standards of work from the expedition's seamen.

At times the sailors' disciplined conduct was central to scientific work. In the *Terra Nova* expedition's Northern Party, one of the main daily scientific tasks was the collection of two-hourly meteorological observations, which provided a record of temperature, pressure, and wind patterns.[93] Regular observations were increasingly important within meteorological fieldwork, providing domestic scientists with the data needed to construct climate models. At first the task of collecting these was carried out by the scientific staff, but they struggled to maintain these regular observations. Raymond Priestley reports that the two cooks, Able Seaman Harry Dickason and Petty Officer Frank Browning, "learned to read the thermometers and barometers accurately" and took over the role.[94] In contrast to the scientists, the two men missed very few observations.[95] The same qualities that made a successful seaman, such as self-control and timekeeping, were transferable to this meteorological work. The role of working-class naval ratings in such work stands in contrast to existing literature that focuses on the importance of naval officers in nineteenth-century meteorology.[96]

Focusing on the self-discipline of expedition members calls into question the idea of expeditions as being fully under the control of expedition leaders. This shift is positive, as it helps us to see the contributions and agency of the labor forces they employed. But we should not lose sight of the fact that not all expeditions ran smoothly.

PUNISHMENTS AND VIOLENCE

Many explorers used violence to crush subaltern resistance. Stanley inflicted a string of violent punishments on those under his command and on local people who failed to provide sufficient assistance to his expedition. Reports of such actions (often taken from Stanley's own writings) caused significant problems for him in Britain, leading to an inquiry into his conduct.[97] Stanley more than deserves his poor reputation. In *How I Found Livingstone* (1872), he boasted that he was "the first white man who had taken a slave-chain with him on his travels" and used it on those who sought to leave his expedition. He also bragged about how he personally struck African men with a donkey whip and threatened them with a loaded gun, portraying such actions as central to the apparent success of his journey.[98] Stanley was more enthusiastic about dishing out violent punishments than many other explorers. But even those with better reputations engaged in similar practices. David Livingstone, who in Britain was remembered for his kind, patient attitude toward those he commanded, relied on violence on several occasions. On his final expedition, he personally gave a group of Sepoys "some smart cuts with a cane" after they gave away their loads.[99]

Violence against racialized people was commonplace on expeditions in Africa. Sometimes it reflected an expedition leader's loss of self-control. In one more reflective passage, Thomson reported that explorers' books often contained passages about the "troublesome conduct, desertions, obstinacy" of porters. "But we are never told how much they have to bear from their masters, when they have lost all moral control of themselves by incessant troubles, and throw, on the most frivolous pretexts, sticks, books, boots, or the strongest language of Billingsgate, at them."[100]

Other travelers boasted about the fear they apparently inspired among those they commanded. Burton claimed that the EAE's Nyamwezi porters wrote a song about him that described him as "Muzungu mbaya," meaning the "wicked white man." Burton wore this insult as a badge of honor, declaring that a "good white man" would have been "flayed without flinching" in East Africa. Burton's nickname was more likely a reflection of caravan customs of joking, as Burton was probably too unwell to be a genuinely authoritarian commander.[101] In his published narrative, Burton sought to minimize the weakness and dependence he felt while traveling.

But seeming tough was not the only pressure. Many explorers tried to show themselves as less cruel than Arab caravan leaders who operated in the area. In *Journal of the Discovery of the Source of the Nile*, Speke reported that his porters were deliberately going slower than they ought to. He claimed he was "reluctant to apply the stick, as the Arabs would have done."[102] Cameron commented that, "when travelling with Arabs, I found that we had treated our men with too much consideration."[103] As I will show, it is not clear that European leaders were genuinely less cruel than Arab commanders, but such comments served an important rhetorical function, in that they allowed explorers to write about violence while at the same time emphasizing the humaneness of their leadership. They also ensured that any violence that explorers did mention did not appear gratuitous in the eyes of readers.

However they depicted their expeditions, we should not lose sight of the real violence and pain that such punishments entailed. In *A Walk across Africa* (1864), Grant claimed that "discipline had to be upheld in several instances by inflicting corporeal punishment" and described a disturbing incident where one porter "roared for mercy during his flogging" after being accused of stealing cloth to buy food.[104] Speke's private letters show his reliance on force in clearer terms than his published account.[105] He wrote to the secretary of the RGS describing how his headmen brought "thieves" to him for punishment: "At first I awarded fifty lashes, but that not having the desired effect, increased to 100, and finally to 150, which is the most I have dared to give for fear of mortal consequences."[106] At one point on his 1860–63 expedition, Speke lost his temper explosively, punching his headman Sidi Mubarak Bombay so hard that it left a scar on his forehead, still visible years later.[107]

Burton also ordered floggings for disobedience and says he showed himself "ready to enforce obedience by any means and every means."[108] The use of corporeal punishment against Black bodies was representative of a broader trend within Victorian racism. While the mid-nineteenth century saw the "demise in the use of flogging against 'whites' within the British Empire," its use against colonized people continued for much longer. Flogging was justified, in part, by portraying Black skin as less sensitive to pain and non-Europeans as incapable of being controlled or disciplined by nonviolent means. Whipping was a punishment that "participated in the making of racialized subjects" in explorers' accounts of travel.[109]

Explorers did not always have free reign in how they treated their subordinates. In East Africa, for instance, the ability to desert illustrated how limited explorers' powers could be. Explorers often acknowledged that in the East African interior, porters could (and did) "run away at any moment."[110] This factor clearly limited the ability of an expedition leader to command through coercion alone. Speke claimed that the Waungwana men of his party often retorted when he chastised them: "Flog me if you like, but don't keep count against me, else I shall run away; and what will you do then?"[111] This pressure kept some explorers from committing more extreme acts of violence.

But later travelers did engage in punitive violence and even murder. Mary Kingsley shot her revolver at a group of porters who deserted her on Mount Cameroon, but she fortunately did not hit any.[112] Halford Mackinder is often lauded as the founder of modern geography. On an expedition to the summit of Mount Kenya 1899, he drove his Zanzibari porters forward by threatening them with his rifle, shooting off rounds when they threatened or tried to leave. Eight porters were "shot by orders" on the journey, but the archival records do not record exactly when and where this happened. The execution of these porters was omitted in all published accounts of the journey.[113] The level of violence employed by Mackinder was extreme but not unique.

The rear column of the Emin Pasha Relief Expedition (1886–89) saw comparable acts committed under the leadership of Major Edmund Barttelot (fig. 20). He and the other officers ordered and oversaw numerous acts of violence and murder. This included the execution of one Sudanese man named Burgari, who tried to desert after having been flogged, taking some goat meat to eat while escaping.[114] The unfortunate man was brought back by men under the command of an Arab trader in the area, who asked the explorers not to shoot him, as they did not want "the sin of his death" to fall upon them.[115] Unmoved by this appeal, the man was tied to a stake and shot by ten of his comrades the following morning. This act was not isolated. Barttelot committed other violent acts—for instance, personally kicking a boy so hard that he died of a resulting infection. He and the other officers also oversaw regular floggings of expedition members and acts of violence against local people.[116]

The death of John Henry, an East African convert to Christianity who had briefly visited London, illustrates the extreme levels of violence carried out of the rear columns' officers. Henry was accused of desertion and with the theft of a revolver by Barttelot.[117] He was publicly sentenced to death by Barttelot in front of the rest of the party. According to William Bonny, another officer on the expedition, Barttelot was apparently planning to show clemency to the man and punish him in some other way.[118] But when the death sentence was announced, the other Zanzibaris vigorously objected, stating that, while they thought the man deserved some punishment, they would all desert if he was shot. After this display of resistance, Barttelot felt that any act of mercy would be seen as a sign of weakness and was desperate to either shoot Henry or to cut off his ear. Another officer thought the man should be flogged rather than shot and "had rows continuously" with Barttelot over the question. Barttelot eventually ordered that the man receive three hundred lashes with a "stout stick"—almost double the number Speke ever dared to give. Henry stood through this appalling punishment without making a noise. But his wounds soon became infected, and he died two days later. Barttelot's rampage of violence was only stopped when he was himself shot by an African man who was concerned that the major was about to attack his wife.[119]

FIG. 20. Major Barttelot as depicted in Stanley's *In Darkest Africa*. Wikimedia Commons.

Such incidents are illustrative of the forms of violence that some British explorers considered themselves entitled to inflict on Black people in the closing decades of the nineteenth century. The fact that they felt able to carry out such punishments shows how growing colonial intervention in the continent shifted the power relations on individual expeditions, enabling explorers to use force in increasingly extreme ways.

It is easy to vilify Barttelot as an individual. But, even at the time, organizations like the Aborigines Protection Society saw the responsibility for such acts of violence as resting with a far broader array of individuals and institutions. The expedition had, after all, been partially funded by the

British government, and many acts of violence were carried out under the watch of the expedition's ultimate leader, Stanley.[120] Another officer on the expedition suggested that much of the responsibility lay with Stanley, as he had appointed Barttelot "absolute master" despite "knowing what a temper" he had "and knowing the major's hatred of the blacks."[121]

Racism was clearly a central motivation for explorers' violence against Africans. Antarctic explorers were more reluctant to commit violence against white bodies and often emphasized the positive culture of cooperation on expeditions in their published accounts. But, even here, violence was not totally out of the picture. On the BNAE, Scott had the cook, Henry Brett, "locked in irons for eight hours" for his failure to follow orders and cook hygienically.[122] He, like other explorers, felt entitled to inflict some limited forms of violence on the bodies of subordinate expedition members. Even Shackleton, who today is widely lauded for his leadership skills, had to resort to threats of force on the Imperial Trans-Antarctic Expedition. After the expedition's ship sunk, the party tried to travel over sea ice to find rescue. Carpenter Harry McNish complained about this work, suggesting that he was no longer obliged to follow orders since the boat had sunk. Some reports suggest that Shackleton threatened to shoot him.[123] Violence—or, in some situations, the threat of it—was a technique that many explorers used.

CONCLUSION

Expeditions could be profoundly violent, even murderous, undertakings. Many explorers left lasting marks on the bodies of the people they commanded. Sometimes they even killed. Exploration did not just depend on the bodies of others; it could also involve inflicting violence against them. The degree to which explorers used violence not only depended on their personal temperaments and prejudices (though these were undoubtedly important) but also reflected shifting power relations on expeditions (which themselves might change over the course of a single journey). Travelers in Africa became more violent as European empires expanded into the interior of the continent in the 1880s and 1890s. More must be done to recognize the extreme acts of physical violence that exploration could involve. But, as I have argued, we should also avoid writing a history of exploration in which we ignore the agency, discipline, and self-control of subordinate expedition members.

Discussions of this issue thus present a significant challenge for historians seeking to draw attention to subaltern agency and contributions. Explorers used discussions of this issue to manage their dependence on others—and to impose a singular narrative on the multiple bodies of an expeditionary party. Writing about discipline became a means by which expedition leaders retrospectively claimed that their motivations for a journey were the primary ones. Describing subordinate expedition members as lacking in self-control or motivation (at least without the firm, paternalistic leadership of the European leader) allowed explorers at the same time to acknowledge and minimize the contributions of others. Such descriptions are often laced with racial or class prejudice, portraying African, Asian, and working-class bodies as inherently lacking in foresight and self-control.

While tropes around authority, heroism, and leadership are prevalent in explorers' accounts, a close and contextualized reading of public and private sources suggests that in some contexts explorers had to accommodate subaltern agency. In practice discipline on most expeditions was very different from Peary's portrayal of the expedition leader as the "head" of an expedition. Explorers had to regulate their own behavior to fit in with preexisting customs and conventions and often relied on the authority of intermediaries. At the same time, leaders often depended on the self-control of subalterns to a degree that often goes unappreciated. Focusing on the ways bodies were disciplined thus allows our discussions to move beyond simply listing subaltern contributions and to foreground something of their agency within the history of exploration.

4

ANIMAL BODIES

The exhibition on the Victorian traveler and naturalist Frank Oates in Gilbert White's House and Gardens in Hampshire, England, exemplifies some of the challenges of writing animals into the history of exploration. The exhibit includes a display on Oates's two loyal dogs, Rock and Rail. When Oates died in Matabeleland in 1875, Rail reportedly returned "many miles" to the graveside of his former owner, apparently to stand guard as the "sole mourner." In both the display and in Oates's posthumously published book, the incident is reported as a touching story of human-animal companionship and canine loyalty. But, moving as this story may at first glance appear, a reflection on its context renders it more disturbing. In such depictions the dogs are described in more sympathetic language than any of the African men who formed part of his expeditionary party. In Oates's book the men are called "boys," and their "troublesome" behavior is imputed to bring on the "relapse" of fever that killed him. The "faithful" dog is used to illustrate the alleged disloyalty of the African men.[1]

Other writers also used animals to devalue subaltern humans. In an unpublished letter, penned in East Africa, to the secretary of the Royal Geographical Society (RGS), John Hanning Speke reported on the progress of his 1860–63 journey. "The numerical losses of the expedition in animal life superior and inferior amounted," he wrote, "to a total of 168 individuals."[2] Speke then went on to break down this number. He included, disturbingly, 128 Nyamwezi porters and 6 Cape Riflemen, as "animal" losses to the party. He portrayed the entire logistical workforce of the expedition—including African men—as subhuman. As these two examples show, writing animals into the history of exploration raises fundamental questions about the relative value given to different contributions and experiences within travelers' accounts.

Until fairly recently historians of science and geographical fieldwork often gave "peripheral attention" to the role of nonhumans "as objects of study, companions, pests or danger."[3] But there is a growing literature on this subject.[4] We now understand that most expeditions were "hybrid" undertakings, relying on both human and animal bodies.[5] The increased attention to animals on expeditions reflects broader developments in both the humanities and social sciences.[6] Geographers have devoted growing attention to the spatial dimensions of human-animal relations.[7] There is also more literature on animals and empire. Such works have argued that expanding the "concept of the subaltern" to include animals can deepen our understanding of both nineteenth century science and imperialism.[8] My approach also reflects recent developments in labor history and the growing awareness that humans and animals frequently "co-labor" together.[9]

Scholars interested in contemporary injustices have demonstrated the value of studying histories of human oppression and animality together.[10] The animalization of women and people of color play a central role in broader discourses of patriarchy and racism.[11] Historically, too, the links between racism and discussions of animals mean that studying the two together can be insightful. Ideas about racial difference "operated alongside, within, and against questions of animal difference and human-animal difference."[12] Such ideas were often based on scientific theories (not generally accepted even at the time) that questioned the idea that there was one singular species of humanity. Such "polygenist" thinkers claimed that different "races" represented different human species that had either developed separately or had become distinct by the process of evolution.[13]

Animals could be explicitly used to uphold racist ideologies—for instance, through the naming of animals. On Robert F. Scott's British National Antarctic Expedition (BNAE), the lead dog was "a black dog with some tawny markings" and was named after the n–word.[14] Aware of the connotations of this name, Scott claimed that it "wholly failed to capture the grandeur of his nature."[15] Similarly, on Scott's final expedition and Ernest Shackleton's (1907–9) *Nimrod* expedition, the explorers both chose to name one of their ponies after the ethnic slur "Chinaman."[16] As this shows, even in Antarctica, a continent without an indigenous population, exploration was still conducted in ways that reinforced an opposition between white explorers and racialized others. Animals played a key part in this process.

Given this context writing animal bodies into the history of exploration does present some quite fundamental challenges because our knowledge about them is entirely mediated by humans, as animal studies scholars have examined.[17] There are also more specific challenges. Explorers generally viewed animal bodies as resources, and their writings often tell us more about their own views and experiences than they do about those of the animals.[18] But there are glimmers of animal agency and curiosity that shine through even the relatively unpromising sources sometimes available. While it is not possible to fully understand animals' experiences, it is possible to write an "animal sensitive" history of exploration.[19] I argue that writing such a history can deepen our understanding of exploration, and the ways in which subaltern contributions were minimized within dominant accounts.

HUNTING AND VIOLENCE

Dramatic encounters with carnivorous animals made great stories of adventure. One of the most famous of such incidents is David Livingstone's "encounter" with a lion, famously reported in *Missionary Travels* (1857). When working as a missionary in Mabotsa (now in South Africa), Livingstone had gone to help the local people kill a lion that had been attacking their livestock (fig. 21). After shooting the animal, it leapt at him, biting his shoulder, and knocking him to the ground. "He shook me as a terrier dog does a rat," Livingstone wrote. He was saved by another man who fired at the animal, distracting it at this crucial moment. The lion bit this man in his thigh before dropping dead from the shots he had received.

Livingstone portrayed the incident in heroic terms, emphasizing that local people declared the animal "to be the largest lion that they had ever seen." He also claimed to feel little pain during the attack and reported his wounds in detached medical language.[20] Such dramatic encounters with dangerous animals were an important part of a daring travel narrative. But it is worth remembering that, however he portrayed the incident in his account, it left lasting marks on his body, scarring his skin and limiting the strength and mobility of his shoulder and arm.

FIG. 21. David Livingstone attacked by a lion, 1874. Courtesy of the Wellcome Collection. Image ref: 561253i.

Even when no such incidents occurred, carnivorous animals were often used to emphasize the dangers that explorers faced. As Mary Kingsley quipped:

> Now a crocodile drifting down in deep water, or lying asleep with its jaws open on a sand-bank in the sun, is a picturesque adornment to the landscape when you are on the deck of a steamer, and you can write home about it and frighten your relations on your behalf; but when you are away among the swamps in a small dug-out canoe, and that crocodile and his relations are awake. . . . He is highly interesting, and you may not be able to write home about him—and you get frightened on your own behalf. For crocodiles can, and often do, in such places, grab at people in small canoes.[21]

Encounters with carnivorous animals might be the most memorable incidents in explorers' accounts. But some of the most dangerous creatures that explorers encountered were mosquitoes, which fed on travelers' blood and spread malaria and other tropical diseases. Some travelers took these dangers seriously, while others dismissed local warnings.

While traveling in the horn of Africa, Richard Burton was informed by Somali people that mosquitoes spread "deadly fevers." He dismissed this information as "superstition," which, he claimed "probably arises from the fact that mosquitos and fevers become formidable about the same time."[22] For much of the nineteenth century, British travelers saw fevers as caused by miasmic emanations from rotting vegetable matter and dismissed other factors. In general, though, developments in the newly emerging field of tropical medicine meant that explorers did become more aware of the risks of insect bites, developing a variety of protective measures, including mosquito nets, avoiding travel in dawn and dusk, and taking prophylactics such as quinine to help combat the effects of malaria. The 1883 edition of *Hints to Travellers* advised that "in the tropics, or in malarious regions, always sleep under mosquito curtains," claiming that the "advantages derived from strictly attending to this rule cannot be exaggerated."[23] Livingstone was enthralled with the mosquito net, writing that "the invention deserves a statue in Westminster Abbey."[24]

Mosquitos were not the only small creatures that explorers' encountered. Burton complained that while staying in one hut in East Africa, his room was "populated with pigeons, rats, and flies by day, and with mosquitos, bugs, and fleas, by night."[25] Sometimes flies could be more than an annoyance. When traveling through West African mangrove swamps, Kingsley wrote of a "big silent mangrove-fly which lays an egg in you under the skins; the egg becomes a maggot and stays there until it feels fit to enter external life."[26] Beetles could also cause problems. Speke reported that on the East African Expedition (EAE), he lit a candle in the middle of the night and was besieged by hundreds of beetles. With his efforts to brush these insects away proving futile, he tried to get some sleep, only to be awoken by a beetle crawling into his ear. His efforts to remove the insect only pushed it further down the canal, and it began to dig away at the inside of Speke's ear "like a rabbit at a hole." Speke tried to remove the creature using tobacco, oil, and salt with no success. He eventually killed it with a pen knife, but injured his ear in the process, leading to some of the sensory problems discussed elsewhere in this book.[27]

Some small animals were even more threatening. In many regions poisonous snakes were a particular danger. After traveling in a dry area of East Africa, Joseph Thomson reported that small venomous snakes were "very common . . . and we came frequently into unpleasant proximity. More than once did they glide over my foot, and I have frequently had my presence of mind upset by finding myself sitting down beside them." In his book Thomson used these animals to try and portray himself as braver than local people. He claimed that he once "scattered a whole village, and my men besides, by coming amongst them holding by the neck a large green snake, eight feet long, which I had stunned and then picked up."[28] After traveling to Brazil, the naturalist and later RGS president Henry Walter Bates wrote in similar terms, claiming that local people told him that another venomous snake, the jararaca, "attacks man without provocation." But Bates was never attacked by the jararaca, even though he nearly stepped on it several times. The disparity, he claimed, showed that, on certain subjects "the testimony of the natives of a wild country is utterly worthless."[29] While explorers often drew on standard colonial tropes to present themselves as less afraid of snakes than local people, many clearly remained cautious around them. For instance, after stopping to camp one day, Thomson spotted a "large green snake" in a tree overhead. He tried to bring it down, but with no success, and dryly noted that he had to sleep "with the pleasant consciousness of danger in our immediate neighbourhood."[30]

Explorers were much more comfortable when they were the hunters rather than prey. Many explorers killed animals for no other reason than "sport."[31] Hunting became "a central part of imperial culture" in the nineteenth century and was linked to ideas about elite British masculinity.[32] Not only was killing animals a chance to display bravery, bushcraft skills and marksmanship, but a carnivorous diet also became a way in which many travelers expressed their masculinity.[33]

The British explorer Frederick Jackson was a particular enthusiast for hunting. He led an the Jackson-Harmsworth Expedition (JHE, 1894–97) to the Archipelago of Franz Josef Land, high in the Russian Arctic. The expedition was tasked with carrying out a survey of the area, but Jackson spent much of the time hunting polar bears and walruses. In its first year, the expedition killed nearly sixty polar bears, more than half of which were shot by the expedition's leader.[34] Hunting was a means to provide the expedition with fresh meat, which Jackson correctly thought would help prevent scurvy. But he also enjoyed hunting for its own sake and included long detailed descriptions of his trips in his published account. Some of these incidents contain elements of danger and adventure. One polar bear succeeded in attacking Jackson. "I could feel his warm breath on upon my face, could see the gleam of his teeth and the shape of his long grey tongue and the furious glare in his savage eyes," he reported.[35] In this case Jackson managed to free himself by thrusting his rifle into the bear's mouth, and the bear, too, escaped. Most of the expedition's animal encounters were not so dramatic. The party's main challenge was the fact that their weapons often merely injured animals. Not only was this cruel, but it also often meant their bodies could not be recovered for food, rendering their suffering futile.

Franz Josef Land was unpopulated. In other places European hunting methods could have negative effects on Indigenous people. Bates discusses different approaches toward turtle hunting adopted by Indigenous people. Amazonian people hunted turtles using an arrow with a strong lance attached to a piece of twine. Several arrows would be shot into a turtle, which would then be pulled toward the hunter. After watching this practice Bates noted that "the men had shot about a score of nearly full-grown turtles."[36] Bates's settler guide, in contrast, used nets to capture animals, a practice that the Indigenous people did not consider "lawful sport." Nets resulted in a much higher yield—eighty animals in twenty minutes—but was a far less sustainable practice, as most of the animals captured were young and may not have had a chance to breed.[37]

Unsustainable hunting practices by European explorers may have had dire consequences for Indigenous peoples. Inuit oral histories of the Franklin Expedition suggest that the presence of the explorers' ships near King William Island coincided with a period of acute famine for the Inuit population, which contributed to the almost complete depopulation of the island. While HMS *Erebus* and HMS *Terror* were stuck in the ice nearby, Inuit reported that they "could not kill anything for their food. . . . They (the Innuits) really believed that the presence of the Koblunas (whites) in that part of the country was the cause of all their trouble."[38] Periodic famines were not unusual among Indigenous Arctic people, and the appearance of the two ships does seem to have coincided with a period of particularly bad weather. But the arrival of more than one hundred explorers, armed with modern weaponry and desperate to secure fresh meat to prevent scurvy, would have put significant strain on food resources in the area. This theory is made particularly plausible because other Inuit histories also suggest that Franklin's men may have carried out a large caribou hunt on the Western side of King William Island.[39]

African explorers also killed numerous animals. Speke was a keen hunter, and his book *Journal of the Discovery of the Source of the Nile* (1863) includes an appendix with a "list of large game bagged by the East African Expedition" including rhinos, zebra, buffalos, hippos, antelopes, and giraffes.[40] His account is replete with shooting narratives. Yet, compared with hunting in other contexts, the explorers felt that their efforts were rather modest. James A. Grant commented that "the country we have passed through does not abound with game, like South Africa . . . so we have not been murdering game as they do there."[41] The explorers took every opportunity for hunting that they could. Speke shot at animals even when he had little prospect of recovering their bodies as food or zoological specimens.[42]

Samuel Baker, like Speke, had a similar love of hunting, and his watercolor paintings from his Nile expedition depict these activities in lurid detail. One image shows Baker "bagging giraffes at a long shot, about 100 yards." Another depicts him shooting a lion at "close quarters."[43] Unlike many other explorers, Baker was more explicit in writing about the role of hunting in his identity as a British gentleman. In a book on his hunting in Ceylon (now Sri Lanka), he complained that the British love of hunting "cannot be understood by many foreigners. . . . He hunts for the pot; and by Englishmen alone is the glorious feeling shared of true, fair, and manly sport." Baker claimed that he had never met a true "sportsman" who was not a "straightforward honorable man," and that real sportsmen shunned animal cruelty.[44] Hunting on expeditions was about far more than providing food: it could also be about the display of an upper-class masculinity.

But in Africa, as in the Arctic, hunting served important practical purposes. The progress of many expeditions was made possible by the killing and consumption of animal bodies. Hunting for the pot also reduced the quantity of trade goods that explorers needed to carry and reduced travelers' dependence on local rulers, who otherwise provided explorers with goats, chickens, and cows in return for cloth, beads, or wire.

Hunting was particularly vital if explorers traveled through areas of shortage and famine. In one letter to the secretary of the RGS, written from East Africa, Speke reported the expedition's problems with buying grain due to local shortages: "Already we had begun to live on buffalo and zebra flesh and what few jungle leaves, herbs or fruits we could pick up."[45]

His companion Grant described the same period in similar terms: "Our supplies of grain frequently ran out in camp, but the sportsman need never starve in the country we passed through."[46] Larger animals could provide a significant calorific boost to a hungry party. Henry Morton Stanley, for instance, reported shooting a giraffe that yielded some 993 lbs. of meat.[47] But not all animals were equally prized. Grant wrote that zebra "smelt of a stable, caused instant disgust and nausea." Buffalo meat was, in contrast, "as fine as that which any English butcher can produce."[48]

Meat was "an important component of or substitute for wages to African labor," so hunting could be a means of keeping morale up in a party.[49] As noted, many of the men and women on East African expeditions were Muslims and would not, therefore, eat animals they considered haram, or that had not been correctly slaughtered. Muslim expedition members, Speke claimed, were sometimes "too fastidious" to eat meat that had been incorrectly slaughtered or prepared.[50] Halal hunting involved quickly cutting the throat of any animal that did not die immediately after being shot, so as to reduce their suffering, and saying "Bismillah." Stanley reported that his Muslim gun-bearers often carried out this duty, rushing to dispatch any injured animals.[51] Europeans did not have total control over hunting and had to accommodate themselves to African agency.

The necessity of killing an animal quickly perhaps partly explains Stanley's complaints about the inadequacies of his guns. On his first expedition to Africa, he grumbled that the weapons that he brought "were unfit for African game," as the animals he shot were often only injured. "What is wanted for this country," he claimed, "is the real bone-crusher—that will drop every animal shot in its tracks."[52] In part such passages reflect the relationship between weapons, hunting, and European masculinity. But, given the importance of killing hunted animals quickly to Muslim expedition members, Stanley's comments might also be a sign that he sought to ensure animals killed were halal.

Explorers also used accounts of shooting to disparage the bravery of African people. Hunting was seldom a solitary process, and explorers often carried it out with African men who carried supplies and weapons and directed them to appropriate locations. In their published narratives, explorers often used dangerous encounters with animals to portray themselves as braver than African men. Speke reported being charged by a wild bull that he was hunting with two African men (referred to in his

FIG. 22. "The wounded boar comes to a halt!" from Stanley's *How I found Livingstone* illustrates how hunting tales were used to portray African men as cowardly. Courtesy of the Wellcome Collection.

accounts as "boys") named Suliman and Farj. He wrote that he bravely stood his ground in the face of the animal's aggression and managed to shoot the animal dead at the last moment. He claimed that Suliman, "with the instinct of a monkey, made a violent spring and swung himself by a bough immediately over the beast, whilst Farj bolted away."[53] Speke dehumanizes Suliman by suggesting that his desire to avoid being charged by a bull is somehow animalistic.[54] Stanley included a similar incident in *How I Found Livingstone* (1872; see fig. 22).

Not all explorers found joy in hunting. In Antarctica expeditions also relied on local animals for food. But Scott always saw it as a necessary evil. Scott was far from a vegetarian: his expeditions brought with them huge quantities of tinned meat and live sheep from New Zealand that were slaughtered and frozen as the explorers entered the southern ice. From the outset Scott saw hunting locally caught seals and penguins as a legitimate food source. Yet he was always saddened that one of his expedition's actions was to bring death to polar animals. "It seemed a terrible desecration to come to this quiet spot only to murder its innocent inhabitants and stain the white snow with blood," he wrote afterward, "but necessities are often hideous, and man must live."[55] Here we see

FIG. 23. Explorers with a captured emperor penguin, British Antarctic Expedition, 1911. Reproduced with permission of the Scott Polar Research Institute, University of Cambridge. Image ref: P2005/5/1135.

Scott valuing animal life quite differently. He had few qualms about killing domesticated sheep but was more troubled by wild polar animals. This difference in attitude was perhaps a reflection of broader portrayals of Antarctica as a pure, virginal space. But Scott may also have felt that killing relatively helpless seals and penguins (who had no experience of land-based predators and did not, therefore, flee) was far less heroic than hunters in East Africa killing larger and more agile animals (see fig. 23).

Scott's more nuanced approach to animal agency is illustrated by another incident. In his 1911 diary, Scott described one "extraordinary scene" in which a group of six or seven orcas attacked the expedition's photographer Hebert Ponting. Ponting was taking photographs on a slab of sea ice (with two dogs standing nearby) when the whales attacked. The orcas dived and swam underneath, striking in a coordinated way:

> The next moment the whole floe under him and the dogs heaved up and split into fragments. One could hear the "booming" noise as the whales rose under the ice and struck it with their backs. Whale after whale rose under the ice, setting it rocking fiercely; luckily Ponting kept his feet and was able to fly to security. By an extraor-

dinary chance also, the splits had been made around and between the dogs, so that neither of them fell into the water.[56]

The attack on Ponting is clearly an example of the coordinated hunting strategy of some Antarctic orcas, who have been documented using similar tactics to hunt seals. Such incidents have demonstrated the intelligence sociability of these animals. These qualities are clearly reflected in Scott's descriptions of the incident. He noted, following Ponting's escape, that "it was clear that the whales shared our astonishment, for one after another their huge hideous heads shot vertically into the air through the cracks which they had made. . . . There cannot be a doubt that they looked up to see what had happened to Ponting and the dogs."[57] While the explorers were afraid of being eaten, they found the inquisitiveness and intellect of the orcas disturbing in a quite different way.

Scott and Ponting depicted the incident in different terms. Ponting described the orcas as "the devils of the sea" and the whale that showed a particular interest in him as a "brute." He portrayed the creatures' actions as a manifestation of their inhuman (and almost monstrous) aggression, showing little curiosity about the animals themselves. In contrast Scott was amazed by the "deliberate cunning," of the orcas and the fact that "they could act in unison." He wrote: "It is clear that they are endowed with singular intelligence, and in future we shall treat that intelligence with every respect."[58]

But Scott sought to regain control of the situation in another way. In the aftermath he wrote a passage in his diary titled "Notes on the Killer or Grampus (*Orca Gladiator)*" outlining his scientific knowledge about orcas (possibly researched from books in the expedition's library). The passage describes the orcas' bodies in scientific terms, noting their average size, number of teeth, and habits.[59] Scott sought to emphasize his position as an observer rather than the observed, subjecting the attacking whales to zoological scrutiny.

ZOOLOGY

Travel was intimately connected to nineteenth-century zoology. Charles Darwin's journey on the *Beagle*, for instance, played an important role in the development of his broader theory of evolution.[60] He had been inspired by Alexander von Humboldt's travels and writings, which changed scientific attitudes toward the natural world in the nineteenth century.[61] For some travelers, such as Bates, zoology was a primary motivation for their journeys. But even nonexperts became involved in the collection of natural history specimens—often the dead bodies or skeletons of newly "discovered" animals, which were then circulated through networks of interested people in Europe and the United States.[62] Explorers were also encouraged to sketch any new animals they encountered—even if specimens had been collected, as these could deteriorate considerably on their journeys.

Like botany and anthropology, zoology was a branch of science that women explorers were more able engage with than geography. Soon after its founding, the Zoological Society of London allowed the admission of women on the same basis as men, and women made up between 4 and 7 percent of the society's fellowship throughout the nineteenth century. The RGS, in contrast, did not open its fellowship to women until 1913, though twenty-two women were admitted in 1892–93.[63] Women travelers often emphasized their zoological motivations in their published accounts. In *Travels in West Africa* (1897), Kingsley claimed that her journey had been motivated by a search for "beetles, fishes, and fetish." Conforming to common literary tropes about feminine modesty, Kingsley downplayed the importance of her collections, though she brought several species to the attention of Western scientists.[64]

Even explorers who wrote about their zoological achievements in less self-effacing terms might struggle to convince domestic scientists that the creatures they saw in the field existed. This was in part because collecting and preserving specimens was not a straightforward task: conditions in the field often rendered preserving animals difficult, and the specimens brought back did not always satisfy the requirements of domestic audiences.[65]

These challenges are illustrated by the search for gorilla specimens in the 1850 and 1860s. At the time the scientific community became fascinated by evidence of gorillas in western equatorial Africa. Explorers searched the region to try and bring back a preserved gorilla corpse as definitive proof of their existence. In 1861 the American Paul Du Chaillu arrived in London with stuffed gorillas and skulls and exhibited them at the RGS and the Royal Institution. But even with such seemingly conclusive evidence in hand, armchair scientists did not fully accept his findings because of broader inconsistencies in his accounts.[66]

Du Chaillu's problems show how hard it could be to convince domestic scientists.

Burton's subsequent trip to the region gives us some insight into how difficult finding, collecting, and preserving such specimens could be. In 1862–63 Burton traveled to Gabon in an effort "to shoot a specimen or specimens of the gorilla, and, if possible, to buy or catch a youngster."[67] Burton relied heavily on African labor, in particular a man called Fortuene. Burton noted that he had "the frank and open countenance of a sportsman" and was "a man of note in his tribe, as a hunter always is."[68] Fortuene directed the efforts to find a gorilla and recruited more men to help with the hunt. Burton grew frustrated that they "utterly failed" in killing, capturing, or purchasing a specimen.[69] Fortuene did eventually get hold of the body of a chimpanzee, which was apparently sent by his "bushmen." While the animal was not the one he had set out to find, Burton measured the body, and Fortuene and his men then tried to preserve it for transport to England. But before a week had passed "decomposition set in," and further efforts were made to try and preserve it. The remains were eventually sent to Britain, but they suffered from poor preservation. Burton later visited the animal, which had been reassembled and displayed, but he was unimpressed: "The colour had changed, and the broad-chested, square-framed, pot-bellied, and portly old bully-boy of the woods had become a wretched pigeon-breasted, lean-flanked, shrunk-limbed, hungry-looking beggar."[70]

In response to such issues, domestic institutions sought to standardize such practices of collection and preservation. Travel guides, such as the RGS's *Hints to Travellers* therefore offered direction on "how best to profit by their opportunities of benefiting science by collecting new or rare species, and how to preserve and safely transmit their specimens."[71] Another strategy was to capture and transport living animals. Bates kept several animals as pets while traveling in the Amazon, including several primates, a toucan, and a parrot. Many of these animals suffered during their captivity. One monkey was killed by another in a dispute over food, another by a neighbor's dog, and a further primate died of a lung disease within only three weeks of capture.[72] Bates' experiences were not unique. The JHE captured several polar bear cubs, hoping to bring the animals back for exhibition in the London Zoo. They fed the infant animals with improvised teats and managed to keep them alive while in the Arctic, but all died on the voyage back to England.[73]

Studying zoological work can be used to surface the skills and contributions of African men. Local people often had a far greater knowledge of how, when, and where to capture animals than expedition leaders. Frank Oates's description of how he came to acquire the body of two African Grey Hornbills in Matabeleland is fairly typical of how such contributions were reported by explorers. "The boys," he wrote, "brought me a young hornbill, and I was taken to the nest." The nest was in a hollow high up in a tree, so capturing the birds presumably involved climbing the tree and killing the animals. Once there they—Oates does not say who—killed and captured an older hen. "I kept both birds," Oates wrote. But he complained that those with him "poked out and pulled the wing feathers off the old hen when I was not looking," presumably for their own use.[74] This incident is revealing. Oates uses the patronizing term "boys" to refer to the adult African men with whom he traveled. These men seem to have done all the actual work of finding and killing the hornbills. Despite this Oates felt that they had no claim to any part of its body and was affronted by the fact that others in his party were also collecting feathers for their own uses.[75]

Other explorers were more willing to acknowledge the skilled Indigenous labor that underpinned their zoological collecting. Many animals collected by Bates in the Amazon were killed by Indigenous people with a blowpipe. He found that it was "a far more useful weapon in the forest than a gun, for the report of a firearm alarms the whole flock of birds or monkeys feeding on a tree, whilst the silent poisoned dart brings the animals down one by one until the sportsman has a heap of slain by his side." Bates was unable to use the weapon himself, writing that, while he had some lessons in using the instrument, he found it "very difficult to hold steady the long tubes." He was also impressed by the skilled labor that went into the production of this weapon, noting that they were made from two separate pieces of wood, which were scooped out with an animal's tooth to form each half of the tube. "To do this with the necessary accuracy requires an enormous amount of patient labour, and considerable mechanical ability," he wrote.[76] The animals explorers searched for and killed while conducting zoological work were the objects of attention because, to the explorers, they appeared exotic and novel. But we should also remember that expeditions often included animals, which they used as sources of labor, companionship, and food. It is to these often more "ordinary" animals that I now turn.[77]

ANIMAL EXPEDITION MEMBERS

Some animals found their way onto expeditions against the wishes of expeditions leaders. Rats and mice were a ubiquitous feature of ships in the nineteenth and early twentieth century.[78] These animals could cause problems for the humans on board, as they would often eat clothes, food, and other items. Maritime explorers often brought ships cats with them to try to control the spread of such creatures. As we have already seen, animals could be a vital part of an expedition's workforce. Explorers' relationships with these creatures are no less complex than with those they encountered. Animal bodies were often used to try and increase the range of an expedition, sometimes using the specific biophysical traits of a particular animal to overcome challenging terrain. In the run-up to World War I, some explorers did experiment with motor cars and tractors, but such efforts were not generally successful. In this era explorers relied heavily on animal bodies. Their dependence on animals might lead them to humanize and personify certain creatures. But, at the same time, animals were generally treated as disposable resources, and their suffering and death is often portrayed as the inevitable price of a successful expedition. While there were general patterns, these relationships took different forms in various environmental contexts and in relation to different animal species.

Ponies, mules, horses, and donkeys were some of the most common expeditionary animals. It is not hard to understand why explorers sought to use these creatures, as they greatly extended both the range of an expeditionary party and the quantity of goods that could be transported. Few relied on horses or ponies more than Isabella Bird. She suffered from a long long-standing spinal condition, which rendered walking long distances painful. While roaming in remote areas, she rode cross-saddle, with her feet on either side of the pony, rather than side-saddle, which was generally considered far more feminine.[79] Riding allowed Bird to move in ways that were impossible when relying on her own body and was itself a form of physical therapy for Bird that often left her stronger than being stationary. When traveling across Persia in 1890, Bird reported that she was "very far from well in Tehran but the gallop over the desert down here did me great good."[80] More recent scientific research has demonstrated that horse riding can be an effective therapy for people with disabilities, due to the motion.[81] Riding enabled Bird to increase her bodily capacity, demonstrating relations of dependence that go beyond mere transport, but blur the boundaries between the bodies of horse and rider.

Bird depicts such animals in various levels of detail. She viewed some animals as "companions."[82] In *A Lady's Life in the Rocky Mountains* (1879), Bird described "Birdie," the bay Indian pony she rode for much of her trip through Colorado, as "a little beauty, with legs of iron, fast, enduring, gentle and wise." Throughout their time together, she referred to Birdie by name and became attached to her, describing how she was amused by "her funny ways."[83] In these passages we see the emotional and affective relationship between Bird and the pony—a bond that Bird considered worthy of recording.

On her journey to lesser Tibet, Bird rode a horse named Gyalpo. She described him as "a beautiful creature, Badakshani bred, of Arab blood, a silver-grey, as light as a greyhound and as strong as a carthorse. He was higher in the scale of intellect than any horse of my acquaintance."[84] Bird's description of this horse is like that of her Rocky Mountain pony. But Bird also commented on the importance of blood and breed in the apparent intelligence of this animal. As recent scholarship has shown, these concepts assumed new importance in the Victorian period, developing in direct relationship to scientific racism and eugenics.[85] Within such discourses, "pure blooded" animals were portrayed as better behaved and more intelligent than those of mixed parentage. Similar arguments were made by racist European scientists and explorers about humans of mixed heritage. Gyalpo was named and humanized within Bird's travel account, while many of the humans on which she relied were (as argued in chapters 2 and 3) anonymized in her published narrative. Bird's description of this animal thus draws our attention to the complex logics affecting the ways humans and animals were valued.

Other ponies and horses get only a passing mention. In *Unbeaten Tracks in Japan* (1880), Bird discussed "the Japanese pack-horse" in general, noting that she had "ridden, or rather sat, upon seventy-six horses, all horrible."[86] In another work she devoted a chapter to "the Korean pony," again, describing such animals as a group rather than as individuals.[87] She reported that these animals "resented every attempt at friendliness with both teeth and heels."[88] The degree to which the contributions of different animals are visible depended not only on their work and breed but also on the degree to which Bird formed relationships with them.

While Bird was able to use horses and ponies effectively, similar animals suffered severely when British explorers tried to use them in environmental contexts to which they were unsuited. In East Africa British explorers used donkeys, mules, and sometimes horses on their journeys (fig. 24). These animals were ridden by the explorers, used to "carry the sick" as well as to transport baggage and other equipment.[89] Equids, the explorers also hoped, might help to reduce their dependence on local labor.

These animals had a high mortality rate, because of endemic diseases spread by the tsetse fly.[90] Such maladies had horrifying effects on the animals, causing their bodies to swell, and liquids to ooze out.[91] Speke reported that most of the mules and donkeys brought on his 1860–63 expedition soon died from "distemper."[92] Insects were not the only threat. Larger predators also hunted expeditionary animals. Burton reported that the EAE lost three of its asses to East African hunting dogs. "In all cases they were attacked by night with a loud wrangling shriek, and the piece of flesh was raggedly torn from the hind quarter," Burton noted. Donkeys purchased in the interior (rather than mules transported from Zanzibar) seemed better able to fend off attack and disease.[93] One donkey bought on the EAE accompanied Speke on his later expedition with Grant. The animal therefore became a "great pet" of Speke's, and he was aggrieved when the animal (by this time pregnant) was killed in a dispute with local rulers.[94] Even more hardy animals often met a grisly end.

Using such animals was not always easy for the African men who cared for these them. Thomson gave an evocative description of the difficulties of driving donkeys:

> I began to learn what a dreadful nuisance the ordinary Unyamwesi donkey is. We had five of them bringing up the rear, each with a load equal to what would be carried by two men. Each one, however, required the services of a man to tug, swear at, and thrash him. So that a donkey was only equal after all to one man. But their value for purposes of transport sank into insignificance when one thought of

FIG. 24. "Explorers in East Africa" from Burton's *The Lake Regions* shows the interdependence of human and animal labor. Courtesy of the New York public Library. Image ref: 1248347.

> the immense amount of trouble and delay they caused to the entire caravan. In a week the donkey-boys had lost all moral control of themselves, and indulged in nothing but profane language, till it was feared they would become insane from the amount of irritation they were subjected to.[95]

As this quote shows, it is not easy to separate human and animal labor on expeditions. Explorers often directly compared the contributions of the donkeys to those of the porters. Both the donkeys and those employed to drive them seem to have suffered severely from this unpleasant and unsuccessful form of multispecies collaboration. Thomson suggests that proximity with the donkeys had a corrosive effect on the moral standing of the men who drove them, perhaps even dehumanizing them in his eyes.

Attempts to use ponies on polar expeditions, pioneered on the JHE, were little more successful. In 1893 Jackson traveled across the Russian Arctic and into Lapland to learn survival and transport techniques from Nenets and Sami people indigenous to these areas in preparation for an expedition to Franz Josef Land. Ponies were always a second choice. While in the Arctic, he had been impressed by the way that Indigenous people were able to travel long distances on sleds pulled by reindeer, commenting that the "reindeer is the camel of the Arctic desert."[96] But Jackson could not find a way to transport appropriate food for the animals to Franz Josef Land and instead decided to use "extremely hardy" Siberian ponies on his expedition.[97] Jackson was impressed with the performance of the ponies of the JHE (though none returned alive) and recommended that other polar explorers follow his lead in the use of these animals.

During his 1907–9 expedition, Shackleton saw Manchurian ponies as his main hope of getting to the Pole, after hearing of the "good work" similar animals did on Jackson's expedition.[98] Scott also followed this lead and took nineteen Siberian ponies with him on his final expedition (fig. 25) and recruited seven "Indian Transport Mules" from Shimla in northern India in case his first attempt on the pole failed.[99] The explorers often wrote positively about the performance of these animals on their expeditions. But, in reality, ponies are ill suited for Antarctic travel in quite fundamental ways: their bodies would sweat, and the sweat froze on their skin; their hooves sank into the soft snow; and their stomachs could only digest bulky food, which increased the loads explorers had to carry.[100] The long maritime journeys the animals had to make before the start of an Antarctic expedition also weakened them before they even arrived in the continent. Scott also had problems because many of his ponies were old and unhealthy.[101]

Because ponies suffered so much in Antarctica, explorers had to go to great lengths to protect their bodies from the environment: building a specially heated stable for them, cooking them hot food, exercising them daily during the polar winter to strengthen their muscles, making special clothes to keep them warm, and constructing snow walls at the end of a day's march to protect their bodies from the wind overnight.[102] Both officers and men played a significant role in training and caring for the ponies—activities that might be understood as acts of affection and companionship. Apsley Cherry-Garrard commented that the sailors had a "profound" love of animals.[103] But these acts of care cannot be separated from the importance of the ponies in Scott's South Pole plans. Caring for the animals was also a question of success and survival, ensuring that they could be exploited for labor and food as the explorers traveled south. When the animals were killed, Scott reported, men in all the tents were "thoroughly enjoying" the ponies' flesh, suggesting that these relation-

FIG. 25. Captain Oates and some of the ponies on board the *Terra Nova*, 1910. Three dogs are also visible in the foreground. Reproduced with permission of the Scott Polar Research Institute P2005/5/1354.

ships of affection had clear limits.[104] None of the ponies or mules taken on these expeditions survived their trips to Antarctica.

Despite their preference for horses and ponies, British explorers did also try to use the bodies of specially adapted animals to facilitate their journeys. Both Scott and Shackleton used Siberian huskies on their expeditions, with varying degrees of success. These dogs are well adapted for life in cold environments: they are light, allowing them to walk over the snow bridges; have thick coats; can hydrate themselves with snow (saving fuel); and pant rather than sweat when hot and on the march, meaning (unlike ponies and humans) they can travel in much lower temperatures without getting too cold themselves.[105] They require less food than a person to pull the same load.[106] If led effectively, as by Roald Amundsen on his 1910–12 South Pole expedition, they allowed the humans to travel with comparative ease, although at a considerable cost to the canines.[107]

From the outset some British explorers had ethical qualms about using dogs in such a way. Before the departure of the BNAE, the second-in-command, Albert Armitage, and Clements Markham, the president of the RGS, wrote to each other sharing their concerns about the "cruelty of dog driving."[108] Scott's experiences on the expedition did nothing to dispel these fears about the ethics of using canine labor. On the expedition's farthest-south journey, Scott took nineteen dogs with him.[109] Suffering from bad food and poor handling, the dogs soon began to die and suffer.[110] As the party traveled, the dogs deteriorated until they were "too weak to walk alongside the sledges" and were then "despatched" by Edward Wilson.[111] Scott initially hoped some animals might pull through, but all were eventually killed. This experience had an important impact on Scott's understanding of the cost of relying on canines.

Scott commented on the hardship that dogs experienced, arguing that this "did rob sledge travelling of much of its glory."[112] His reluctance to use these animals was partly shaped by changing understandings of masculinity and heroism discussed in chapter 1.[113] Overall Scott argued that using dogs was advantageous and could be ethical, provided the animals were well cared for, but did have moral qualms about using (and killing) them.[114]

The problems with the dogs on the BNAE demonstrate how relations of dependence were two-directional. The men relied on the dogs for transport, but the animals required appropriate care from the explorers. When such relations broke down, the animals suffered and died, while Scott was left with a guilty conscience. Scott's limitations on this front were even commented on at the time. In a 1906 review of Scott's book *The Voyage of the* Discovery (1905), the American polar explorer and geographer Angelo Heilprin argued that "with a sustained dog-team . . . a much farther point south would have been attained than that officially recorded and with a much less expenditure of bodily force" and compared the journey unfavorably to the efforts of the American explorer Robert E. Peary in Greenland.[115] Amundsen commented on the problems British explorers had with huskies, asking: "Can it be that the dog has not understood his master? Or is it the master who has not understood his dog?"[116]

These discussions reflected broader social attitudes. This was an era in which the exploitation of animals for both food and transport was intensified through changes in diet and the expansion of horse transport in cities. But the period also saw new debates about people's ethical obligations to animals, illustrated by the emergence of vegetarian and anti-vivisectionist movements.[117] Explorers' experiences of animal agency sometimes took these discussions into interesting new directions.

While Scott could reconcile himself to depending on canine bodies, he struggled to come to terms with the disciplinary structures within huskie packs.[118] In *The Voyage of the* Discovery, he describes his shock at the egalitarian but hierarchical nature of dog society and the hostility of the group to any dog that received special treatment from the men. He was horrified by such violent forms of discipline, noting his distress when the pack of dogs killed two of their companions who had been treated differently by the men. He responded by punishing the "murderers"; "one by one," he wrote, "they were led out and severely chastised in front of their victims." This punishment had little effect, and Scott conceded that "the dogs evidently didn't know what it was all about. You can't change dog nature."[119]

Unable to enforce their own standards of discipline on the dogs, the explorers worked with the dogs' hierarchical structures. Each group had one leader who, Scott wrote, "is unceasingly watchful, he never pauses to parley, but attacks at the first sign of insolence, for he knows well that the sharpest and quickest fang commands the situation."[120] Scott reported that they had to put the alpha dog at the front while sledging, as "he chose his place naturally as the leader, and if put in any other position would make himself so unpleasant to his neighbors, and generally behave so ill, that he was very quickly shifted."[121] Scott also claimed that this lead dog would enforce discipline among the others, acting as an intermediary of sorts. The humans relied on canine forms of regulation, and the success of their journey depended on their ability to adapt to and work with the structures of the dog pack. But Scott continued to have reservations about the violent behavior of some of the dogs.[122]

It was not just in the polar regions that explorers depended on the bodies of specially adapted animals. In lesser Tibet Bird's party found their bodies affected by altitude sickness. "Horses and mules," she reported, "are unable to carry their loads, and men suffer from vertigo, vomiting, violent headache, and bleeding from the nose, mouth, and ears, as well of prostrations of strength." Several yaks were used to move the expedition's supplies across high, rugged mountain passes. Bird also rode a yak, writing that it had an "uncertain temper, and is not favorably disposed towards his rider."[123] The bodies of yaks are suited to life at high altitudes, with large hearts and lungs that enable them to breathe the thin air, adapted circulatory systems, and a high metabolism.[124] The animals were also able to cross difficult terrain, and Bird noted that a yak can "carry loads where it might be supposed that only a goat could climb."[125] These animals therefore enabled the party to travel to areas it would have otherwise proved impossible to reach.

In Australia, too, explorers sought to use specially adapted animal bodies to carry them and their supplies across the country's interior. In the first half of the nineteenth century, explorers used horses, ponies, and bullocks, but, as elsewhere, the animals had high mortality rates, as they required water daily, which was not always easy to secure in arid areas.[126] Camels seemed to provide the answer, as they could travel three times as far without water and carry larger loads.[127] Robert O'Hara Burke and William John Wills were the first to use camels extensively to transport the expedition's supplies across the continent. The twenty-six camels they used had diverse origins, ranging from Rajasthan, Baluchistan, and Yemen.[128] Because the animals could carry large supplies of water, explorers were therefore less beholden to local geography and relied less on Aboriginal guides than many other Australian explorers. These factors ultimately contributed to their demise at Cooper's Creek in June 1861, as Burke overestimated the camels' abilities when planning his return journey.[129]

Later explorers of Australia used camels more successfully by employing of South Asian cameleers to care for them. The men—often referred to as "Afghans," even if they came from other parts of the Indian subcontinent—knew how to care for the camels, but generally received less pay than white expedition members and were sometimes poorly treated by expedition leaders.[130] Nevertheless, cameleers were often an indispensable part of an expeditions labor force. On the Elder Scientific Exploring Expedition (1891–92), which sought to map Southwest Australia, the expedition's five cameleers—Hadji Shah Mahomet, Mahyedin, Alumgool, Mahmoud Azim, and Abdul—cared for the expedition's forty-four camels.[131] Alongside a litany of other activities, this involved collecting water, making and repairing saddles, and finding camels that had become lost. Despite the important work of such men, expedition leaders often disparaged their contributions. In his published account, David Lindsay, the expedition's leader, complained about his difficulties managing the cameleers and claimed that they used too much water to wash their clothes.[132] The cameleers were Muslims and would not eat the flesh of animals that had been incorrectly slaughtered. Lindsay found their autonomy difficult to come to terms with and moaned that it meant that he "cannot control their meat."[133] But these men were so central to the success of the expedition that Lindsay could not proceed without them. As Dane Kennedy has examined, the expedition's white scientists grew resentful of the way that Lindsay treated the cameleers, leading to conflicts that eventually brought the expedition to a "premature end."[134] Using unfamiliar animals might help explorers to overcome logistical problems, but it often involved hiring autonomous human labor forces to ensure their bodies were appropriately cared for.

Focusing on the work of caring for animals can foreground the role of individuals underappreciated in traditional accounts of exploration. When Cecil Meares, the *Terra Nova* expedition's chief dog handler, was in Siberia buying ponies and dogs, he recruited two experienced men from within the Russian Empire to care for the huskies and ponies on their journey from Siberia to New Zealand (where they joined the expedition). These two men were Anton Omelchenko, a groom from Ukraine, and Dmitriy Girov, a dog handler from an island in the far east of Russia.[135] Cherry-Garrard reported that "they proved such good fellows and so useful" that they joined the expedition and traveled to Antarctica.[136] The British often described these men in condescending terms, referring to them as "boys," and they often receive scant attention in subsequent histories.[137] Yet Scott's diary also shows they played a key role in caring for the expedition's animals.[138]

In other contexts, too, focusing on the labor of caring for animals can draw attention to the work of specialist and skilled laborers. In *Korea and Her Neighbours* (1898), Bird noted that each pony in her party was accompanied by a guide called a "mapu." These men and women walked with the ponies, loaded them, and even cooked the ponies a hot meal of beans and grains three times a day, doing so with "great care." Bird wrote that the guides were "very anxious" about the welfare of the animals "and take very great care of them, seeing to what passes as their comfort before their own."[139] These comments demonstrate the hybridity of expeditionary parties. They also show something of the labor and experiences of people we know little about from a biographical perspective. But such descriptions are tantalizingly brief, and we are left without a detailed understanding of the experiences of the ponies or the people who cared for them.

Explorers' writings about the divide between human and animal were often shaped by nineteenth-century racism. Taken together such writings also show the difficulties of centering animal contributions and experiences. Any history of exploration that fails to address animal bodies is fundamentally incomplete. Nonhumans were central to exploration in numerous ways, as a source of labor, companionship, food, clothing, and zoological specimens. They could be threats or impediments to the successful progress of a journey. Many of the challenges of exploration came from the complex interaction of microbes, environments, and animals. This history also exposes the two-sided way that expedition leaders approached animals: their suffering and death factored into the plans of most expedition leaders. But these explorers also humanized and personified certain animals, rendering them more visible than many subaltern people in their parties. In some cases explorers wrote about animals in ways that racialized African and Asian humans. Such descriptions often tell us more about the prejudices of the explorers than they do about the experiences or contributions of those they depended on.

Despite these challenges writing about human-animal relations on expeditions can be a useful strategy to reflect on the techniques by which some explorers dehumanized the people on whom they relied. It is also impossible to recover the contributions of subaltern people to expeditions without reflecting on their relationship to, labor with, and knowledge about animal bodies. As we have seen, often expeditions involved the collaboration of humans and nonhumans, and the development of affective and caring relationships between different species. Such activities were part of the essential work of exploration. This labor was often carried out by subaltern people with specialist knowledge and skills. More exploitative activities, such as hunting and collecting, additionally expose the ways that African, Asian, and Indigenous people contributed through their knowledge and experience. Sometimes they expose the differences between Indigenous and settler mindsets in relation to animals.

While we must acknowledge the importance of animals within the history of exploration, we cannot totally overcome the one-sided sources available. Explorers' accounts often include descriptions of animals only because of their interactions with and usefulness to humans. We cannot know how animals felt about their involuntary membership of the expeditionary party. Yet, as I have shown, we can find rare glimpses of animal agency and curiosity in the archives of exploration. Focusing on even these brief incidents and the role of subaltern people in explorers' natural history collecting throws into question one of the central myths at the heart of exploration: it helps to dispel the idea that whatever the contributions of others to an expedition, knowledge production was something carried out only (or even primarily) by white explorers. I now go on to further investigate the collaborative, collective, and multidirectional nature of knowledge production.

5

SENSING BODIES

The "gaze" of the explorer is often treated as an expression of broader imperial power relations.[1] In her influential book on nineteenth-century travel writing, Mary Louise Pratt portrays the explorer as a "seeing-man . . . whose imperial eyes passively look out and possess." The historian David Arnold argues that the explorer's gaze functioned as an "ordering, even disciplining, mechanism that edited as well as elided information" about the landscapes through which it traveled.[2] These approaches have shaped how we understand the history of geography. The geographer Gillian Rose draws on Pratt's conception of explorers as "seeing-men," suggesting that "their panoramas" were geography's "founding fields."[3] While insightful and nuanced works on the subject have emerged in recent years, exploration is still regularly discussed through reference to views and visions.[4]

These ideas remain powerful because they echo the ways that explorers were instructed to observe. The 1893 edition of the Royal Geographical Society (RGS) advice manual *Hints to Travellers* advised that "the first and best instruments are the traveller's own eyes. Use them constantly, and record your observations on the spot."[5] In reality explorers often struggled with their vision. For instance, when traveling through Australia, John McDouall Stuart complained that "the mirage is so powerful that little bushes appear like great gum-trees, which makes it very difficult to judge what is before us; it is almost as bad as travelling in the dark."[6] The contradiction between Stuart's views and the advice given in *Hints to Travellers* suggests that exploration raised profound question about the reliability of the senses.

In this chapter I argue that producing knowledge on expeditions was complex, relying on multiple senses and multiple bodies. I do so by studying observations in the field on a granular level (and through focusing on a somewhat narrower range of expeditions than previous chapters) and build on recent scholarship that has devoted attention to ways that nineteenth-century geographers relied on locally gathered oral information.[7] To this end I argue that, despite the emphasis on the vision of a solitary traveler within many guidebooks, Victorian and Edwardian geography relied on multiple senses and numerous bodies. Subaltern people played a variety of roles—translating, directing observations, enabling leaders to navigate new environments, and understand what they were looking at. The scientific work of an expedition was not conducted by white explorers using their eyes and ears alone. Geographical research depended on collective and embodied labor.

THE SENSES IN NINETEENTH-CENTURY GEOGRAPHY

The cultures and conventions around geographical exploration were shaped by the long-standing valorization of vision within Western systems of knowledge. On the surface the importance attached to sight can be viewed as a product of the specific biological capabilities of the human body. The eye can, in general, gather information from a greater range than other sensory organs, allowing a much broader area to be surveyed and helping to instill explorers' observations with a level of detached objectivity. The optic nerve is some eighteen times more sensitive than the cochlear nerve of the ear and "is able to transfer an astonishing amount of information to the brain and at a rate of assimilation far greater than that of any other sense organ."[8] Vision can be deliberately directed in ways that hearing, smell, and taste cannot.[9]

That said, we should question the idea that individual senses are "autonomous, with each organ responsible for its own perceptual task."[10] In day-to-day encounters with the world, people use multiple senses at the same time, and sensory experiences are also shaped by memories and emotions (as phenomenological approaches have shown).[11] Contemporary geographical scholarship has done much to critique the idea of landscape as an inert, passive surface separate from the travelers' body.[12]

But there are specific histories to consider as well. Vision was particularly important within nineteenth- and twentieth-century geography because of social and cultural developments brought about by modernization and the Enlightenment. Dramatic changes in technology, government, trade, migration, and culture in the nineteenth century transformed sensory experiences and perceptions for the British public.[13]

Visual representations of the world became increasingly important in popular, scientific, and political cultures.[14] As a result they were increasingly seen as "the only reliable way of knowing it."[15] Victorian culture more broadly saw "expanding the visual field" as an important undertaking.[16] The development of glass windows and lighting technology made sight an increasingly important sense within the Victorian city.[17] The period additionally saw a growing interest in panoramic views and the use of maps to transform "multisensory environments into visual icons."[18]

Historically there were vigorous efforts to define geography as a visual discipline.[19] In 1832, two years after the founding of the RGS, the English topographer and diplomat Martin Leake read a paper before the society on the mapping of Africa's rivers. He claimed that "in Barbarous Africa, nothing short of the ocular inquiries of educated men is sufficient to procure the requisite facts." At the same time, the importance of the ears was downplayed. Leake claimed that "the best collectors of oral information . . . have failed as to arriving at the truth as to the origin, course, and termination of the rivers."[20] Seeing something with one's own eyes thus became an increasingly important standard for establishing scientific truth, marking a departure from earlier travelers such as Alexander von Humboldt, who wrote openly about using his body as a scientific instrument.[21] Increasingly new geographical knowledge "was created through the production of reliable images," including sketches and later photographs.[22]

Victorians were also acutely aware that not all images were trustworthy, leading to vigorous debates about the credibility of different claims. These discussions could raise profound questions about the relationship between "subjective experience and universal law" and between travelers' claims and the theorizations of domestic scientists.[23] Nor were the efforts to define geography as a visual discipline totally successful. Victorian explorers and armchair geographers continued to depend on oral information collected from non-Europeans until the nineteenth century—often because direct observations failed to provide sufficient evidence to settle geographical questions.[24] The Admiralty's publication *A Manual of Scientific Enquiry* (1849) advised that a traveler should "carefully confine himself to the description of what he has himself seen, or what he has learned on undoubted authority."[25] As we shall see, whose authority might be considered "undoubted" was a complex question.

To improve the reliability of explorers' observations, growing importance was attached to the use of instruments and photography.[26] It became more important that a traveler could provide "quantitative and physical evidence that could provide independent verification for his or her claims."[27] Surveying was not only about fixing points on a map but also about representing (both visually and in writing) what geographical landmarks lay at these locations.[28] Literary and artistic representations were inherently subjective, relying on an explorers' senses, personality, and memory.

Cameras seemed to provide a way of recording facts while "eliminating problems caused by unreliable human memory, interpretation, or skill."[29] But the reality was more complex. The American explorer Frederick Cook used photos to authenticate his claim to have reached the summit of Mount McKinley in Alaska and the North Pole. After initial enthusiasm Cook's claims were called into question, as were the validity of his photographs. In 1911 one Australian newspaper reflected contemporary opinion when it claimed that Cook was an "imposter" who had "hoodwinked the public."[30] Other photographers edited their images to convey scenes that were difficult to photograph, sometimes splicing together several negatives to form a single, more impactful picture.[31] For domestic viewers it was hard to establish where and when a specific photograph had been taken, which meant that this technology did not resolve the problems of trust, credibility and identity.

Instruments could also prove problematic. For instance, the political situation in nineteenth-century Tibet made it hard for travelers to deploy scientific instruments. In response the Survey of India used the regulated steps of the Pundits (Indian surveyors) as tools of measurement.[32] Kapil Raj argues that the travels of these Asian men differ from wider European practices of exploration in "quite fundamental ways," particularly in how such travels were comprehended and represented.[33] This may be true, but it would be wrong to view these expeditions as an aberration because they used steps as a form of measurement. On the East African Expedition (EAE), Richard Burton and John Hanning Speke tried to use a pedometer set at the "usual marching pace of laden porters" to measure distances.[34] Burton claimed, however, that the instrument proved "worse than useless" because of conditions in the field.[35] While instruments and cameras added new dimensions to debates about credibility, they also showed that technological developments could not override "the most important *and* the most problematic form of knowledge that an explorer had to offer—direct observation."[36]

Scientific (and now discredited) understandings of the body shaped who was seen as being an accurate observer, privileging the sensory capabilities of upper-class white men.[37] The professionalization of geography meant that confidence was invested "in those explorers whose senses had been disciplined by technical, intellectual and moral training."[38] The senses of upper-class white men had long been privileged above those of others, but these ideas were given a new scientific grounding, due to changing understandings of the nervous system in the nineteenth century.[39] Prominent British scientists, writers, and politicians claimed that racialized and working-class white people were less sensitive to pain and other external stimuli, a theory that undermined their status as reliable observers.[40] The Victorian polymath and author of travel manuals Francis Galton argued that acuteness of the senses was a chief "attribute of a higher race."[41] Related pejorative views shaped scientific understandings of women's bodies. Some male scientists contended that while the "female nervous system" was more sensitive, this sensitivity made women liable to "nervous irritability" rather than accurate observation.[42] Even those who praised the "peculiar powers inherent in ladies' eyes" often did so in very gendered terms. The British author Elisabeth Eastlake claimed that women were better at observing "close and lively details." She also asserted that women travelers were better placed to comment on the "home life" of the countries they traveled through.[43] Their opinions could be useful, but were understood to be qualitatively different and, therefore, unlikely to offer insights on broader political matters.[44] In what follows I examine how issues of location, environment, and identity shaped the ways these issues played out on different expeditions.

UNRELIABLE SENSES, COLLABORATION, AND MEDIATION

These broader cultural conventions shaped how explorers wrote about their own senses and observations. In *The Lake Regions of Central Africa* (1860), Burton emphasized the importance of direct observation in his geographical claims. He stated that "it is the explorer's unpleasant duty throughout these lands to doubt everything that has not been subjected to his own eyes."[45] Burton sought to demonstrate that his topographical claims were based on ocular observation and, therefore, of a higher value than other existing accounts of the region's geography. Burton's description of his "discovery" of Lake Tanganyika is exemplary of many of the broader links between geography, vision, and empire. Pratt even describes it as a paradigmatic "monarch-of-all-I survey" scene in which he looks out in an authoritative and dominating way over the landscape below him.[46] Imperialistic rhetoric abounds in Burton's broader description of "discovering" Lake Tanganyika, but a close reading of his published account demonstrates the unreliability of his own senses and his dependence on others.

For a start, during this key moment in the expedition neither Burton nor his companion Speke could see very well. Both suffered from prostrating fevers that affected their ability to observe correctly. In *The Lake Regions of Central Africa*, Burton described how his eyes had proved unreliable when he first looked at Lake Tanganyika: he initially "gazed in dismay; the remains of my blindness, the veil of trees, and a broad ray of sunshine illuminating but one reach of the lake, had shrunk its fair proportions." He had to ask his African colleague Sidi Mubarak Bombay, who was standing beside him, what he was looking at: "I am of the opinion . . . that is *the* water," Bombay responded.[47] Only following Bombay's oral explanation and after advancing a few paces did Burton understand what he saw. Speke's first experience of this lake was even less dominating. He complained that he could see "nothing but mist and glare" at the time.[48] This incident shows the mediated and multisensory nature of expeditionary observation. In it the explorers' gaze is anything but powerful; it is contingent and unreliable. This is not to say that colonial tropes are absent from the explorer's writings about this incident. Burton went on to describe his view of the lake in the "monarch-of-all-I-survey" style demonstrative of his status as a "seeing-man" with the power to evaluate what he is looking at.[49] But, given his poor physical state and limited vision, it is likely that these passages were added some time later.[50]

Remarkably similar issues emerged when Speke traveled without Burton to the lake he later named after Queen Victoria. Once there Speke's weak eyes proved troublesome: he could only see a comparatively small portion of the large body of water and did not have means or time to fully explore it.[51] His view of the lake was partial—not the kind of broad methodological survey that would have established its size, shape, and relationship to the East African River System. Speke's limited view meant that he depended heavily on oral testimony from a local man named Mansur to fill in the gaps. He described Mansur as "the greatest traveller of the place" to emphasize his credibility.[52] But a more granular reading of Speke's account shows how difficult it is to separate what he heard from what he saw. At one point Speke estimates that the southeast edge of the lake lay "at least forty miles distant," an estimate he based upon the view of a group of hills in this distance. Speke claimed that "*had my attention not been drawn to it*, I should probably have overlooked it and thought there was only a sea horizon before me."[53] There were no other Europeans with Speke at this time, so it was clearly an African man—in this case, presumably Mansur—who directed his sight toward the hills in question. As Speke did not speak a local language, Mansur's conversation with him must have been translated several times.[54] Although Speke stressed the importance of his own gaze in his examination of the lake, multiple pairs of eyes and ears were involved in even these seemingly direct observations.

Contemporary debates about the source of the Nile picked up on how illness affected the comparative reliability of the men's claims. These issues were particularly acute for Speke because he made the bold (and fiercely contested) claim to have "discovered" the source of the Nile. In criticizing Speke and his theories, Burton argued that Speke's blindness and deafness "incapacitated him and prevented him from reading, writing, and observing correctly."[55] Burton also reported that, before reaching the lakes, Speke "had a fainting-fit, which strongly resembled a sun-stroke and which seemed to permanently affect his brain."[56] Because sunstroke insanity was a major concern of Victorian tropical medicine, in suggesting his companion had suffered from it Burton implied that he was an unreliable source of geographical information.[57] These accusations stuck. Summarizing the dispute in the 1890s, the *Edinburgh Review* argued that while Speke had subsequently been proved right about the source of the Nile, he had lacked the evidence to prove his claim at the time, partly because "when Speke was at the head of Tanganyika he was purblind."[58] Victorian commentators clearly understood the limited nature of some explorers' senses.

But the problems the explorers experienced were not isolated to such dramatic incidents. Throughout their journey both men were plagued by periodic bouts of fever-induced hallucinations, including "horrid visions" of "animals of grisliest form, hag-like women and men with heads protruding from their breasts."[59] The confusion the men experienced is ever more apparent in their private correspondence. While the party was traveling toward the lake regions, Speke wrote to the president of the RGS complaining that Burton's eyes "become bad," particularly when the pair tried to take lunar observations.[60] Speke sought to emphasize how he had performed the lion's share of the scientific fieldwork. But his vision was little better: he reported that he managed to take "only one" lunar observation at Ujiji and noted, "My eyesight was very bad so that I have much doubts concerning it."[61] In his private letters from the expedition Burton also wrote of the "distressing blindness" he experienced and stated that Speke had "lost his sight so as to be unable to read, write or observe for some time."[62] Speke also complained that the expedition had become confused by the "zigzags" the party had taken when moving across the country.[63] This disorientation meant that their bearing was calculated "by guess work" rather than by "any true knowledge of the direction we had come in."[64]

The explorers had taken a sketchbook and had originally planned to record the scenery and geographical features at regular intervals, the idea being that observations taken on-site were more reliable than those recalled from memory. Speke noted that plant growth was so dense that "sketching was out of the question."[65] Much of the time, they could only see a few meters in any direction, giving them little knowledge of the broader geography of the country. As elsewhere the "imperial eye appears not as transcendent, all-knowing, global, but instead as situated, partial, local."[66]

For much of their day-to-day navigation, Burton and Speke depended on oral information gathered from Arab and African sources.[67] All the maps produced on the EAE mixed "direct observations with substantial data drawn from local informants." In his *Blackwood* map, which supported his claim to have discovered the Nile's source, the arrangement of many of the key geographical features depicted on the map depended "on layers of oral testimony."[68] This reliance is not clear from the map itself, and Speke did not distinguish between areas that he visited and other areas that he heard about.[69] As Lowri Jones argues, Speke's "sketch maps," sent to the RGS as the expedition traveled through East Africa, reveal in far clearer terms than his published maps his use of oral information gathered from local people and African travelers.[70] Burton's maps devoted greater attention to areas where the expedition had visited, but still downplayed his reliance on existing trade routes and secondhand information.[71] Similar trends emerged on Speke's 1860 expedition. The editing of Speke's *Journal of the Discovery of the Source of the Nile* (1863) removed much of his reliance on African oral evidence, making it seem as if he saw phenomena that he had only heard about.[72] Both explorers depended on oral information but did not fully acknowledge this in public.

Even at the time, it was clear that they had relied on such testimony to support their claims. The ensuing "quarrel" between the two men about the source of the Nile led to a vigorous debate about the validity of oral reports as a source of credible geographical information; Adrian Wisnicki has suggested that the debate marked the "death knell" of such information.[73] But Jones has successfully demonstrated that the criticism of Speke focused on the way he had gathered and used this information, as many of his detractors drew on similar sources.[74] Burton claimed that Speke's recording of oral testimony was unlikely to be accurate because he was "wholly ignorant of Arabic [and] was reduced to depend upon 'Bombay,' who spoke an even more debased dialect than his master." In private he claimed that Speke had misunderstood the man's hand gestures.[75] In attacking Speke, Burton draws our attention to the important role that Bombay played as a translator. Likewise, the armchair geographer James MacQueen dismissed many of Speke's claims for being based on "hearsay" but used oral information extensively in his own writing.[76] In focusing on debates over the validity of oral information in general, previous scholarship has also taken for granted the idea that there was a clear-cut distinction between what the explorers heard and what they saw. It is, of course, impossible to know which bits of information came from visual or oral sources.

The EAE was just one incident in the European exploration and colonization of Africa. But in focusing on it in detail we see that the line between direct observation and oral testimony was more blurry than previous scholarship has appreciated. This was not uncommon. Though Isabella Bird was far less physically unwell than Burton and Speke when she traveled through East Asia, she still had her vision directed by others.

The importance of oral testimony in shaping what Bird saw is particularly apparent in her writings about Japan. As noted in chapter 3, she depended heavily on her guide Ito Tsurukichi for translation and oral explanation of the phenomena she was observing. A few remarks of Bird's also expose the degree to which she relied on the senses of others. At one point she attended a Japanese wedding to which Tsurukichi was not invited. Describing the incident she noted the disorientation this caused her: "His absence was like the loss of one of my senses, as I could not get any explanations till [*sic*] afterwards."[77] These comments suggest that much of what Bird reported about Japan came from combining visual observations with Tsurukichi's oral explanations. They also demonstrate that, while he was with her, Bird had become so accustomed to his explanations and direction of her vision that his absence left her confused. Her comments show that her gaze was directed, shaped, and contextualized by information from guides and intermediaries. Tsurukichi's eyes and ears played a central—but not otherwise obvious—role in Bird's other observations throughout the rest of *Unbeaten Tracks in Japan* (1880).

In many of Bird's works, a few brief passages bring to the surface relationships of sensory dependence that took place over the entire course of an expedition. When traveling through Persia, Bird relied on the services of her educated guide and translator Mirza Yusuf. Often his presence and role as a translator went unmentioned in her account. At one point, however, Bird reminded the reader of his presence: "It must be understood that Mirza Yusuf goes with me everywhere as attendant and interpreter."[78] Throughout the rest of the book, the line between information received through (or explained by) Yusuf, and that which Bird observed or obtained directly, is not made clear. In other instances, as in *The Yangtze Valley and Beyond* (1899), Bird further blurred the line between things she saw and information she heard. In a passage describing the Chinese New Year celebrations she noted that

> Be-dien [her guide] told me that in the city, where there are many *literati* and rich men, there were houses with all their woodwork covered with gold-sprinkled red paper, and on the lintels five slips expressing the desire of the owner for the "five blessings": riches, health, love of virtue, longevity, and a natural death. Over some shops was a decorated slip, "May Rich customers enter this door," and in many stately vestibules, in which handsome presentation coffins were reared on end, there were costly scrolls inscribed with aphorisms and other sentences.

In this passage the line between what Bird is told and what she sees directly is unclear. Bird suggested that she saw many of the signs in question, but she elsewhere conceded that she cannot read Chinese characters; thus it is likely that all of the information about the content of the signs must have been explained to her orally.[79] Much of what initially appears as direct observation in her account is mediated through the eyes of Bedien or possibly through subsequent reading. But this process was far from seamless, and she also described the challenge of reconciling her experiences with existing knowledge about the country. She complained that it was "not an easy matter" to even correctly identify the names of many places.[80] Such difficulties led her to "agree with the dictum of Socrates: The body is a hindrance to acquiring knowledge, and sight and hearing are not to be trusted."[81]

Because of broader pejorative understandings of women's sensory abilities, Bird came to use photography as an important tool in her travel narratives and lectures. In some senses the camera acted as an extension of Bird's body, providing a seemingly incontestable record of her experiences.[82] In 1892 and 1893, before traveling to Korea and China, Bird undertook extensive photographic training, including at the RGS.[83] Through her acquisition and use of this new skill, she used photography to highlight the reliability of her own observations and to "bolster her status as a professional travel writer and explorer."[84]

Bird took and developed more than 1,200 pictures when traveling through East Asia in the 1890s and made use of them in her books on the journey.[85] She also drew on photographs and lantern slides as a source of credibility in her 1897 RGS lecture, reflecting the growing acceptance of these technologies within the organization.[86] Bird stressed that her photographs were purely scientific and faithful representations of her experiences. In a letter to her publisher, she claimed that she was able to produce photographs "which are faithful though not artistic records of what I see."[87] We should not take her claims about the lack of artistry in her photographs at face value. Expeditionary photography was neither purely artistic nor scientific but combined features of both.[88]

Nor were Bird's photographs unmediated records of what she saw. In common with many photographers, many of her images were staged and conformed to broader stereotypes about the regions through which she traveled.[89] Anna Stoddart's Edwardian biography includes a "snapshot" of Bird taken in China in 1895 that illustrates the composed nature of her photographs (fig. 26). The picture shows Bird with her camera in front of a group of people posing for a fairly typical "ethnographic" photograph.[90] The image presents her "as both orchestrator and part of the event," throwing into question the idea that her photographs were neutral documents; it also suggests that composing such scenes involved negotiation with local people.[91]

Examining the ways Bird used her camera also draws attention to the contributions of Chinese chair-bearers to her expedition. For much of her journey through China, Bird's camera was stored under the seat of her chair and was carried on the shoulders of Chinese men.[92] The chair-bearers shaped the composition and content of her pictures as well. Bird reported with sarcasm and astonishment that they showed "a growing interest in photography, reaching the extent of pointing out objects at times 'to make pictures of!'"[93] But Bird did not describe their influences in detail, contributing to their broader erasure in her account, though one tantalizing incident suggests that she sometimes heeded their advice. When trying to photograph a temple in Paoning Fu (now Langzhong) in Sichuan, Bird found that "it was not possible to get any point of view on the level, for a photograph." Her chair-bearers stepped in and "suggested my taking one from the stage of an open temple theatre opposite, and brought a ladder to help me up with."[94] Bird's photographic work thus relied on the bodies and senses of numerous other individuals.

So far I have shown the importance of guides and intermediaries in directing explorers' observations. But what did explorers do when they traveled to an unpopulated continent? Broader histories of exploration present Antarctica as a space where white explorers were able to "test their physical and mental mettle against an unforgiving environment without indigenous mediation or interference."[95] In contrast I argue that the lack of oral information and the unreliability of the explorers' senses rendered Antarctica a confusing region to explore. They had to cooperate extensively and to use multiple senses to understand the world around them. The relatively homogeneous makeup of these expeditions meant that Robert F. Scott and other writers were prepared to describe such collaborative work in far greater detail.

When Scott first arrived in Antarctica on the British National Antarctic Expedition (BNAE), he found his senses bamboozled. His eyes proved

FIG. 26. "Snapshot" of Isabella Bird taken in Satow by Mr. Mackenzie. Courtesy of New York public Library. Image ref: 1112385.

unable to establish distance and perspective in a land filled with snow and ice. On traveling past one area of coast on the expedition's ship, the *Discovery*, Scott wrote that

> the land as we approached it looked illusively near; the sky was overcast, and the higher land was hidden in cloud, but beneath this sheet of grey and black rocks stood out with such distinctness that one was wholly deceived as to their distance. So strong was this deception that the engines were eased when we were nearly two miles from the cliffs, under the impression that they were only a few hundred yards away; we only discovered the mistake when we saw a colony of penguins, and found that even with my glasses it was impossible to distinguish individuals.[96]

This phenomenon has been examined by others and is a product of the fact that spaces with few recognizable features confuse the human eye, rendering perceptions of depth and scale inaccurate.[97] Indigenous people in the Arctic with experience of such phenomena helped American and British explorers to navigate in such conditions, but Scott did not draw on such experience.[98] His frequent use of the word "we" is also revealing, in that it suggests that the men overcame their own individual confusion through oral communication between each other.

Visual confusion was no minor matter. It caused Scott to report geographical features incorrectly: "I noted in my diary that there was on our right 'a curious indentation like the crater of a volcano', and this was really the strait between the island and mainland some ten miles across."[99] As this shows Antarctica rendered individual pairs of eyes an unreliable instrument for establishing the size, shape, and nature of geographical features.[100] Accurate observations could only be produced collaboratively. Such teamwork reflects the expedition's naval background, as such observations were "collective and disciplined projects of work, involving the coordination of many hands and eyes."[101] But Scott clearly viewed Antarctica as a particularly confusing space, suggesting that the level of collaboration required in this context was unusual.

Problems with polar mirages had been a long-standing problem for nineteenth-century British and North American polar explorers, leading them to falsely report land where none existed.[102] In Antarctica such mirages are caused by a temperature inversion that serves to bends rays of light and cause objects beyond the horizon to appear much closer than they really are.[103] Scott presented himself as largely able to overcome the sensory confusion of Antarctica by critically interrogating what he saw. He reported several inaccuracies in the maps and observations of previous explorers, which he suggested are the result of the confusing visual effects just described.[104] He claimed that an explorer might overcome such confusion by being "exceedingly cautious in believing even what appears to be the evidence of one's own eyes" and through a close study of previous accounts.[105] In his *Terra Nova* expedition (1910–13) diary, Scott gives some clues as to the means by which he interrogated such phenomena, reflecting on a winter-evening talk by the expedition's chief dog handler, Cecil Meares, in which he described a journey to Tibet. Scott commented that Meares had many good qualities, "but a lack of scientific training causes the acceptance of exaggerated appearances, which so often present themselves to travelers when unfamiliar objects are *first seen*." For Scott the initial appearance of an object is deceptive, and it is only through sustained and deliberate interrogation that an object's true dimension can be measured.[106] Distrust of vision, to him, was one of the defining features that separated a geographical explorer from a more adventurous traveler. Scott did not view vision as a definitive and authoritative sense, but one that requires caution and critical interrogation. His comments stand in contrast to Paul Carter's suggestion that explorers displayed a "gullibility with regard to the evidence of the senses that often borders on the comic."[107]

Nor were explorers naive about the ability of instruments and photography to fully overcome such challenges. Scott commented that "instruments provided no answer" to many of their navigational problems. "We could but guess," he noted.[108] The Antarctic environment also presented challenges for the camera as a tool to record landscape. While photography could be effective around the coast, it was less helpful in the interior; there was, as Kathryn Yusoff has argued, "too much indistinguishable white space that lacked contrast and too much light that caused photographic plates to be overexposed." As a result the explorers' bodies were often included in photographs, as their presence "anchored and gave meaning to the vast expanse of non-human space." Paradoxically, the advent of this new technology meant that the body became central to measuring the Antarctic landscape (fig. 27).[109] The problems of photography in Antarctica are reflected in the importance of Edward Wilson's sketches and paintings in the visual record of Scott's expeditions. Wilson claimed that it was important to draw from direct observation rather than "preconceived notion," and his artworks capture a range of visual effects that the explorers found it impossible to photograph.[110]

Although explorers developed experience at testing the credibility of their visual observations, similar challenges continued to emerge even as the BNAE left Antarctica after two years in the ice. As he departed from the continent, Scott investigated a group of islands previously seen by the polar explorer James Clark Ross and the sealing captain John Balleny in the nineteenth century; in one location Ross had reported three islands, Balleny only one. When the BNAE reached the location of the island(s), the expedition's men became "puzzled when we found that by no means could we reconcile the accounts of the two explorers . . . and, at first, the clouded condition of the land added much to this difficulty."[111] After Scott had "read the accounts many times," he came up with a theory that reconciled both: "Balleny had seen it from the north, in which direction it presents a comparatively narrow front. . . . At a later date, Ross must have seen the same island, and, as we saw was quite possible, from a great

FIG. 27. An advert for tabloid photographic chemicals, using a photograph from the British Antarctic Expedition, illustrates the use of men to create a sense of scale. Wikimedia Commons.

distance he must have imagined it to be divided into three and hence made the mistake of naming it as a separate group."[112] The ability of the expedition to fix the islands accurately was helped by the fact that the cloud cover lifted at the right moment. On the one hand, Scott's solution to this geographical dilemma shows that the party's prolonged stay in Antarctica had enabled it to interrogate the accuracy of what it saw. On the other hand, Scott's ability to solve the problem still demonstrates that the reliability of his eyes depended on collaboration, weather, perspective, and a close and imaginative reading of previous observations. The explorers' vision is embedded in the physical material world; they did not make observations from a "detached vantage point."[113]

Explorers in both the Arctic and Antarctica experienced other sensory problems in polar environments. In the summer the reflective nature of snow and the extended periods of sunshine meant that the men frequently experienced snow blindness, where the eyes become, in effect, sunburned. Not only was snow blindness painful but it also affected the ability of explorers to keep records or make scientific observations.[114]

Explorers wore goggles to help protect their eyes, and, on Scott's final expedition, they even made similar fittings to protect the eyes of the expedition's animals.[115] Some explorers reported wearing these diligently. On the British Arctic Expedition (1875–76), Albert Markham reported that the men only took the snow spectacles off "after the party had halted for the night and had sought the shelter of their tent" but some men still suffered from mild attacks of this problem.[116] In contrast Scott wrote that, at times, he grew "inexpressibly sick" of the goggles, which he found uncomfortable and made it harder to navigate.[117]

Snow blindness remained a painful and widespread problem on Edwardian-era Antarctic expeditions. Despite years of polar experience, Wilson reported that one day he had a "bad attack of snow glare" on the return from the South Pole in 1912; he "could hardly keep a chink of eye open in goggles to see the course."[118] One of the ways the explorers navigated this problem was by taking it in turns to be the "leader" who was responsible for navigating the sledge. Stoker William Lashly demonstrates the collaborative and multisensory nature of such work in one discussion of how he navigated down the Beardmore Glacier with Captain Edward Evans and Petty Officer Thomas Crean: "Crean has become snow-blind to-day through being leader, so I shall have the job to-morrow, as Mr. Evans seems to get blind rather quickly, so if I lead and he directs me from behind we ought to get along pretty well. I hope my eyes keep alright."[119] Lashly's description demonstrates how sledging involved multisensory relations of dependence that changed according to the condition of individual explorers' eyes. At some point or another, most explorers traveled forward blind, reliant on those around them to direct their movements. Even the leader of a party, who was required to have stronger vision, could not navigate alone, relying on oral directions shouted from behind to stay on course in the confusing Antarctic environment. Cooperation became a way to cope with the bewildering sensory experiences of polar travel.

SMELL, HEARING, AND TOUCH

African explorers depended on multiple senses when traveling and observing. Their noses proved particularly useful in describing the climate and healthiness of an area they were moving through. The importance attached to smell was a product of miasmic theories of disease prevalent at the time: between the 1850s and the 1880s, there was widespread support for the idea that "foul smelling emanations" actually caused ill health.[120] Such theories were a feature of broader campaigns by "scientists and medical men" to shape public opinion in favor of sanitary reform in British cities.[121] Medics in India frequently argued that "miasmas or pestilential air" caused diseases, particularly fevers.[122] The nose played an important role within these developments, as it could be used to detect the disease-causing smells.[123]

Many explorers' writings pick up on these ideas. Describing a trip through a mangrove swamp in West Africa, Mary Kingsley claimed there was a "peculiar smell" in some areas, which was "the breath of the malarial mud, laden with fever."[124] Joseph Thomson described one area of East Africa as a "great festering fever pot" and claimed that the ground was "emitting foul gases continually from the rotting vegetation."[125] In another location, Speke reported that "as Miss [Florence] Nightingale would say, you could palpably smell the fever."[126] Henry Morton Stanley wrote in almost identical terms about traveling through a stretch of jungle: "The odour emitted from its fell plants was so rank, so pungently acrid, and the miasma from its decayed vegetation so dense, that I expected every moment to see my-self and men drop down in paroxysms of acute fever."[127]

Burton saw the coastal regions of East Africa as particularly dangerous because of the "miasmic fever" they picked up there, but he also noted similar problems in other areas.[128] When they passed through a region where "the water was bad, and a mortal smell of decay was emitted by the dark, dank ground," Burton complained that the "reeking miasma" led to an attack of fever in Speke.[129] Burton's views about air and smells of the areas he traveled to were not peripheral to the impact of the expedition. Some explorers viewed the "miasmic" air of the tropics as proof that the area was "inimical to human physical and cultural development," and particularly unsuitable for the European body, which was ill adapted to life there.[130] Thomson drew on these ideas explicitly, claiming that the "fever-laden winds" caused the alleged "physical debility and mental prostration of the natives," and expressed his surprise that the inhabitants of this area had "not been swept off the earth entirely."[131] For British travelers how the air smelt and felt were important questions in establishing the suitability of areas for human settlement.

Even to explorers the air of Africa was not universally bad, though. In describing more hospitable regions, the explorers commented on what they smelled and how they felt. Samuel Baker often commented on the quality of the air as he traveled, noting areas where it felt particularly "invigorating" or left him feeling "refreshed."[132] During a "forced halt" in a more salubrious area, Burton reported that he found that the "change of air was the most fitting restorative" to his continuing problems with numbness and fever.[133] On his later expedition with Speke, James Grant found the mountain regions of East Africa particularly pleasant. He noted that, after the rains, the air had a "delightful freshness."[134] Burton described the mountain air of Usagara as a "sanatorium," because it was free from "the heavy exhalations . . . emitted by the decayed vegetation," and had a pleasant breeze and "copious dew which renders the nights peculiarly pleasant to a European." The smell and feel of an area were indications of its healthiness. Burton suggested that "should Europeans ever settle in Eastern Africa as merchants or missionaries, here they might reside until acclimatized for the interior."[135]

David Livingstone was less convinced by the idea that noxious smells were the only vector of disease.[136] But he did view mountainous areas of Africa—such as the Batoka Plateau—as more suitable for European settlement.[137] While explorers' writings might not lead to successful settlements, the idea that hill stations and mountain regions were more healthy places for Europeans was mainstream within the British Empire at the time, shaping patterns of settlement, migration, and governance.[138] In using multiple senses to show that there were areas of East Africa with fresh, nonmiasmic air, the explorers were helping to demonstrate the potential for European habitation and hence helped to pave the way for colonization.

In other environments, too, explorers relied on senses beyond vision. When Henry Walter Bates traveled in Amazonia, he commented that the "multifarious" animals in the evening was one of the things that stuck him most. Each species, he reported, was "sounding its peculiar note," and when frogs and toads chimed in after dark it "created an almost deafening din. This uproar of life, I afterwards found, never wholly ceased, night or day: in course of time I became, like other residents, accustomed to it. . . . After my return to England the death -like stillness of summer days in the country appeared to me as strange as the ringing uproar did on my first arrival at Para."[139] Bates clearly became acclimatized to the sounds of the Amazon, and, throughout his narrative, he reported on the sounds of different animals—often recognizing them without seeing the animal in question. These comments illustrate that while hearing could not be directed in the same way as vision, an explorer could learn to differentiate between an initially overwhelming chorus of noises.

Hearing and touch were important tools for navigation in the dark polar winter. We find particularly insightful descriptions in writings about a march to Cape Crozier to collect emperor penguin eggs conducted by Apsley Cherry-Garrard, Wilson, and Henry "Birdie" Bowers in the Antarctic winter of 1911. The party traveled in near-total darkness. Cherry-Garrard later wrote, "We began to realize now that our eyes were more or less out of action, how much we could do with our feet and ears."[140] Wilson reported a similar experience, noting that "eventually we travelled by ear" using "the feel of the snow under our feet" to tell whether the surface was safe or only a thin bridge over a bottomless crevasse.[141] But it is Cherry-Garrard who describes the newfound importance of touch most evocatively: "The effect of walking in finnesko is much the same as walking in gloves, and you get a sense of touch which nothing else except bare feet could give you. Thus we could feel every small variation in the surface, every crust through which our feet broke, every hardened patch below the soft snow."[142] These incidents foreground the "feeling body" of the explorer and are a far cry from conventional and heroic understandings of polar travel, which generally emphasize visual representations.[143] The reliance on senses beyond vision was partly a result of polar darkness, but it also brought to the fore the often latent reliance of the explorers on these senses in other contexts. Only deprived of their sense of vision did the explorers realize how useful touch and hearing were.

CONCLUSION

Victorian writers and more recent critics of exploration have often referred to explorers as "seeing men," highlighting the importance of their visual observations and the symbolic power of their gaze. This chapter has challenged this idea in several ways. First, it has shown that explorers often struggled to see clearly. For significant periods of their journey through East Africa, Burton and Speke could not see at all because of fever and other tropical diseases; therefore, many of their observations must have been gleaned from oral information and the eyewitness testimony of others. Antarctica, meanwhile, presented distinctive problems, as explorers struggled to see despite the effects, of snow blindness, darkness, and a confusing array of optical effects. The specific causes and duration of these visual impairments varied, shaped by the personal health of different explorers and the specific challenges presented by each environment. In this sense an analysis of the different ways in which explorers approached and struggled with their vision helps to displace the idea that travelers were separate from the landscapes through which they traveled. Their sensory observations were the products of bodies embedded in material environments, which had profound effects on how explorers sensed and experienced the world.[144] Explorers' sensory powers were shaped by local diseases, environmental conditions, but also by their relations with other people.

From these different contexts, describing explorers as "seeing men" ignores the fact that, in the field, they were often blind. Of course, they did often write about their gaze as powerful and drew on colonial tropes, but many of these comments sit at odds with what we know about their physical state while they were traveling, suggesting that they were often later additions aimed at domestic audiences. The motif of the seeing man and the masculine colonial gaze is problematic in other ways when it comes to explorers like Bird and Kinsley. Contemporary writers often suggested that women's eyes were inferior to men's, except perhaps when it came to understanding minor details and domestic affairs. Such women accordingly adopted a variety of techniques to ensure that their findings were credible to domestic audiences, such as, in Bird's case, the use of photography. That said, these travelers still have more in common with masculine explorers than is often acknowledged. Like them they depended on a broad range of guides and intermediaries who are only partially visible in her public and private writings.[145]

Recent critical literature on exploration has begun to address the dependence of Europeans on such oral information but offers only a partial picture. In some senses this approach seeks to replace the explorer as a seeing man by emphasizing their role as a collector and compiler of oral testimony. These approaches have produced some valuable insights, but they have been limited in drawing too clear a line between oral information, direct observation, and other sensory impressions. In reality surviving, navigating, and producing knowledge on an expedition depended on multiple bodies and multiple senses. It is often hard to draw clear distinctions between these various sources of knowledge. Explorers relied heavily on oral information from guides and intermediaries even when making apparently "direct" observations. They often did not know what they were looking at until someone explained it to them. Occasional comments demonstrate that the work of translation and mediation was a central part of expeditionary knowledge production. Unpopulated Antarctica emerges as an exception.

But this only made collaboration more important. The extreme environment produced a strange array of visual effects, and the explorers frequently found themselves confused by what they saw. Neither photography nor instruments could fully overcome these problems. They responded to this situation by depending on multiple senses, by working together, and by comparing what they saw with previous accounts. In this sense, far from demonstrating the triumph of the unmediated European gaze, Antarctic exploration shows their interdependence and their reliance on senses beyond vision, particularly touch. On the one hand, the greater emphasis on collaboration can be understood as part of a broader shift in scientific practices towards more systematic fieldwork. On the other hand, Scott's ability to address the importance of collaboration and cooperation directly shows how questions of identity remained an important consideration in how contributions were valued within published accounts.

The key message of this chapter is, therefore, around the difficulties of drawing clear boundaries around individual senses or individual bodies. Knowledge production on expeditions was a collaborative process. This insight has broader implications in terms of how we think about exploration, demonstrating its collective nature and the difficulties of separating the production of knowledge from other activities such as translation and transportation. The chapter therefore critiques the idea that geographical knowledge was produced only by expedition leaders and makes it possible to understand the multiple individuals involved.

6

EXAMINED BODIES

In 1853 Richard Burton disguised himself as a Pathan dervish and medic in the hope that he might pass unnoticed into the Muslim holy cities of Mecca and Medina. In Burton's eyes the disguise had one particular advantage: pretending to be a religious mystic allowed him to avoid scrutiny of his own body. The dervish, Burton claimed, "may pray or not, marry or remain single as he pleases, be respectable in cloth of frieze as in cloth of gold, and no one asks him—the chartered vagabond—Why comes he here? or Wherefore he goes there?"[1] That Burton, a keen anthropologist, who never missed the chance to examine and describe the bodies of African and Asian people, wanted to avoid such attention tells us something about practices of examination more broadly. Travelers wanted to be able to observe, examine, and classify the bodies of the people they encountered, without having their own bodies examined and observed in a reciprocal way.[2]

These ideas reflected wider cultural conventions in which explorers were encouraged to collect information that would contribute to the nascent discipline of anthropology. Travelers' accounts "provided the basis for the theoretical arguments both of ethnologists and of the social evolutionary writers who succeeded them."[3] Anthropologists and explorers examined the bodies of racialized people—living and dead—in invasive and objectifying ways. From 1883 the Royal Geographical Society's *Hints to Travellers* contained a chapter giving advice on how to take anthropological observations, composed by Edward B. Tylor, the founder of modern cultural anthropology.[4] Even nonexpert travelers were instructed that they could bring back useful knowledge. Tylor claimed that "the characters of men's bodies and minds being matters of common observation, Europeans not specially trained in anthropology, who had happened to be thrown among little-known tribes, often bring home valuable anthropological information." Although it reflected Tylor's interest in cultural difference, the chapter also illustrated how the concept of "race" was central to anthropology in this period. Tylor advised explorers to record non-European people's "characteristic features" through sketches, notes, and photography. Wherever possible travelers were encouraged to measure or photograph the bodies of "a large number (20 to 50) of persons taken indiscriminately as they come" and to collect samples of skeletons, skulls, and hair.[5] These instructions were often carried out.[6] Some explorers also brought living people—in many cases children—to colonial metropoles to be exhibited in front of European and North American publics. Such practices caused the violent objectification of specific bodies, rendering racialized people as "specimens of natural history."[7] Explorers' anthropological works circulated through global networks as well, contributing to destructive ideologies of scientific racism and European notions of cultural superiority.[8]

We should never lose sight of the physical and epistemological violence that explorers' anthropological examinations entailed. But do we miss something if we focus only on the actions of white explorers? Do we reinforce the idea that explorers were the only ones producing new knowledge? In this chapter I argue that we should also devote attention to the various ways that explorers' bodies were examined. This chapter responds to and extends nascent scholarship on this subject; there has been a growing recognition that bodily examination was a two-way process. As Johannes Fabian notes, many explorers objected to African interest, hypocritically viewing it "as an invasion of their privacy" at the same time as they examined the bodies of others.[9] Mary Louise Pratt argues that passages describing travelers' examination by African people are exemplary of "anti-conquest" tropes within such narratives. Explorers, she claims, used these incidents to foreground their own vulnerability and humiliation in the field and to emphasize their benevolent motivations.[10] This may well be the role such incidents played for European audiences. But we should not discount the idea that, in the field, explorers could genuinely find themselves the objects of curiosity, as Dane Kennedy has demonstrated.[11] African people, Kennedy argues, "scrutinized explorers with a frank and relentless determination that stands in challenging contradistinction to the 'imperial gaze' that so many scholars in recent years have attributed to explorers themselves."[12]

There is growing recognition of the two-directional nature of curiosity, but much previous scholarship relies on visual metaphors to explain such encounters. As I will show, travelers were not just the objects of a countergaze; they also found their bodies pinched, poked, stroked, and measured in visceral ways. Erik Mueggler's approach in *The Paper Road* (2011), which builds on Maurice Merleau-Ponty's ideas about vision and the gaze, comes closer to the strategy I employ here.[13] In doing so he highlights how the process of both observing and being observed could make explorers aware of their "deep involvement in the flesh of the world—its viscera, its filth."[14] Through a comparative approach, I push these insights further, using the different ways explorers were observed to show the interactions between them and the political and social contexts through which they traveled. Focusing on the two-way and multisensory process of examination on expeditions shows us that explorers were not the only ones producing knowledge. Examining incidents where they were met with resistance rather than curiosity allows me to interrogate the relationship between exploration and more direct forms of colonial violence.

A study of these issues also sheds light on the relationship between exploration, writing, and power. As I will demonstrate, explorers frequently wrote in disparaging and racist terms when their own bodies were examined by local people, reflecting their frustration at being the observed rather than the observer. Their disparaging writings can—in some cases, at least—be read as an attempt to counteract the powerlessness and dependence they felt while traveling.[15] Narration was instrumental in reasserting control.[16] In reexamining these instances with a critical eye, I refocus attention on the perspectives of local leaders and non-European populations. In addition I discuss the different ways this agency was expressed, including though curiosity and resistance. Such relationships shaped explorers' journeys and the knowledge they were able to produce.

ENCOURAGING AND OBJECTING TO CURIOSITY

In most instances archives and books record intercultural encounters only from explorers' perspectives. The Franklin Expedition, which left England in 1845, is one remarkable exception to this rule, as the only detailed histories of the expedition's final years come from Inuit sources. Following the disappearance of the expedition in the Arctic, several other explorers—notably John Rae, Charles Francis Hall and Knud Rasmussen—gathered Inuit oral testimony about the expedition's fate. The testimony they gathered does present challenges, as the records were shaped by issues of translation, the use of leading questions, and the prejudices of the men who gathered and recorded it. Despite such problems Inuit histories of the Franklin Expedition were central to the efforts of Parks Canada to find the wrecks of HMS *Erebus* and HMS *Terror* in 2014 and 2016, respectively.[17]

The testimonies also offer insights into how Inuit people viewed the explorers' bodies. Those who encountered Franklin's men after they abandoned their ships were shocked at their physical state. One man, Too-shoo-ar-thar-i-u, who happened upon four members of the expedition, reported that one man "looked so bad" that he "could not bear to look at his face," as the explorer's eyes were "all sunk in."[18] Another Inuit man, Mangaq, met some of Franklin's men and reported his interactions with them in similar terms: "We ran up to them, and saw that they were white men. They were thin, with sunken cheeks, like starving men. We took them to our tent and gave them seal meat and blubber. We would gladly have helped them but could not understand their speech."[19] Because of the explorers' limited language skills, these encounters left some Inuit feeling that white men had "quite the same minds as small children. . . . They are easily angered, and when they cannot get their will they are moody and, like children, have the strangest ideas and fancies."[20] Such passages reverse many of the colonial tropes discussed elsewhere in this book, as the explorers are infantilized and rendered objects of Inuit pity and compassion.

After the abandonment of the expedition's icebound ships, many of Franklin's men tried to reach safety by dragging small boats to open water. Some ran out of food on these journeys and resorted to cannibalism. Others died of cold, starvation, and exposure. The bodies of these dead men were often found by Inuit people traveling in the area. Upon finding such bodies, Inuit often noticed the poor physical state of the corpses. One body was memorable to a man named In-nook-poozhee-jook, as it was "very large and tall, and by the state of gums and teeth was terribly sick (bad state.)"[21] The man in question was probably suffering from scurvy when he died. While many explorers died traveling over land or ice, some men returned to the expeditions' ships and died there. These vessels were also explored by Inuit people. One party came across one of the expedition's ships, went inside, and found "many dead men lying in their beds," but were frustrated when the ship sank before they could recover many of the useful items they found on board.[22]

Much attention has been placed on the value of these testimonies in piecing together the fate of the expedition. But this can sometimes lapse into an extractive approach in which Inuit histories are viewed as valuable only for what they tell us about the fate of Europeans. The significance of the encounters for Inuit people should not be ignored. In many cases the details that seemed most important to Inuit people were not those most interesting to Franklin searchers. Inuit accounts often devote attention to the number of knives or useful tools they encountered on or with the arctic explorers. Metal knives were a rare commodity that could materially improve the lives of those in possession of them. The expedition's shipwrecks, abandoned boats, and supplies were an important source of wood and iron, both valuable commodities in the Arctic.

One incident is particularly illustrative of the complex ways that some Inuit people approached the bodies of explorers. Sometime in 1853–54 an Inuit couple came across a "big tent" containing and surrounded by the dead "bones and bodies" of Franklin Expedition members. While her husband was gathering items from around the tent, the woman came across one body, which was half frozen in ice and had a silver chain around his neck with a watch attached to it. To retrieve the watch, the woman chiseled away at the ice with a sharp stone until the body was released, taking care to avoid damaging the cadaver while doing so. The task was unpleasant and memorable, as "besides the tent being filled with frozen corpses—some entire and others mutilated by some of the starving companions, who had cut off much of the flesh with their knives and hatchets and eaten it—this man who had the watch she sought seemed to her to have been the last that died, and his face was just as though he was only asleep." Eventually, she and her husband were able to retrieve the items. Hall reported that "the woman now keeps them very choice, in commemoration of the terrible feelings she had when getting them from the dead Kob-lu-na [white man], whom she dug out of the ice with nothing but a heavy, sharp stone."[23] The woman was both moved and troubled by her close encounter with the dead explorer's body, and sympathetic to the unfortunate fate that befell this fellow human. But she was also prepared to overcome such feelings to retrieve the watch, which was a rare item in this region.

The experiences were recorded by later explorers because the fate of the Franklin Expedition was unknown outside the Arctic. In other contexts we only have explorers' accounts to work with. But even these can be used to tell different stories. Richard Burton and John Hanning Speke's Nile expeditions offer insights into the level of curiosity that explorers encountered as they traveled through East Africa. In their accounts we see that they were the subject of particular attention because of their strange bodies and dress. One of the common themes in their writings is the suggestion that they were a source of amusement and even humor to Africans. In *The Lake Regions of Central Africa* (1860), Burton complained that local people often assembled "to stare, laugh, and wonder at the white man" and that African interest in the explorers was "accompanied with peals of merriment."[24] James Augustus Grant also reported that when he traveled through East Africa with Speke, the men's bodies were the objects of similar curiosity: "The villagers *en route* turned out to see the white men. . . . Of an afternoon, we considered it an extraordinary occurrence if our camp was not thronged by people, curious, well-conducted, some bringing their produce to barter. Women would sit at our tent-doors suckling their infants while cracking jokes at our expense."[25] The travelers may well have stressed the humorous nature of these encounters to avoid having to seriously engage with the motivations of those who examined them. Even if we accept these descriptions as accurate, we get a remarkably different picture of what travel entailed: in the field, travelers might find themselves the objects of scrutiny and the butt of jokes.[26]

Burton's, Grant's, and Speke's experiences seem to have been fairly typical of European explorers in East Africa in the 1860s and 1870s. In *How I Found Livingstone* (1872), Henry Morton Stanley complained that he received so much curious attention when he traveled through Wagogo areas that he felt "not much better than the monkey in the Zoological collection at Central Park, who funny antics elicit such bursts of laughter from young New Yorkers. . . . I philosophically resigned myself in silence for their amusement," he wrote (though he did emphasize his violent reaction to particularly invasive examinations).[27] Near Lake Tanganyika Verney Lovett Cameron claimed that people "turned out in very great numbers, very black and naked, to see what our business might be and seemed very friendly to us notwithstanding their character of universal robbers."[28] Despite being scrutinized himself, Cameron sought to emphasize his role as the observer and evaluator of the people he encountered, commenting on their bodies and behaviors in a disparaging way.

The interest in the travelers' bodies could limit their ability to observe freely. On the East African Expedition, Speke reported, his "French spectacles so excited the crowds of sable gentry who followed the caravan, and they were so boisterously rude, stooping, and peering underneath my wide-awake [hat] to gain a better sight of my double eyes [sunglasses], as they chose to term them, that it became impossible for me to wear them."[29] This quote is particularly revealing, as, without these glasses, Speke himself could not see and had to close his eyes. To overcome his discomfort at finding himself as the object of attention, Speke was prepared to limit his own ability to travel and observe.

When examinations crossed certain lines (and when they felt able to), explorers would try and reassert control. In *What Led to the Discovery of the Source of the Nile* (1864), Speke described an incident at Ugogo "on the highway between the coast and Ujiji." The road "was literally thronged with a legion of black humanity so exasperatingly bold that nothing short of a stick could keep them from jostling me. Poor creatures! They said they had come a long way to see, and now must have a good long stare; for when was there ever a Mzungu [white man] here before?"[30] Speke's comments are both denigrating and revealing: they suggest that he had little objection to being stared at but was prepared to defend himself with force when such bodily examinations took the form of "jostling" rather than a distanced stare. He also sought to reassert control over a situation in which he felt scrutinized and vulnerable by referring to those who had observed him using the animalizing term "creatures." In doing so he sought to recast the incident in terms that reinforced Victorian racial prejudices.

Burton and Speke had quite different reactions to these incidents of examination, which are illustrative of broader tensions produced by such encounters. As an anthropologist Burton vigorously objected to this reversal of positions.[31] He offered his interpretations of the different stares to which he was subject, sometimes ascribing sinister "cannibal" or "pugnacious" feelings behind them.[32] Burton's analysis offers almost no insight into the actual motivations of those looking at him, but it does illustrate how his racial prejudices hardened over the course of the expedition. After he returned he became "one of the most fierce and vocal advocates of the view that black Africans constitute a distinct and inferior species of humankind."[33] Burton used racism to reestablish his position as the observer of others.

Speke, in contrast, had little "ethnographic curiosity" and found African inquisitiveness more amusing than distressing, particularly when it was carried out by men and women of high status (see fig. 28).[34] He began to use this interest in his body to win favor with local rulers. When describing his first trip toward the Victoria Nyanza in 1859, Speke recounted one particularly remarkable incident. A local ruler named Ungugu invited him for dinner and then began to inspect him:

> Squatting by my side, the sultana at once shook hands. Her nimble fingers first manipulated my shoes (the first point of notice in these barefooted climes), then my overalls, then my waistcoat, more particularly the buttons, and then my coat—this latter article being so much admired, that she wished I would present it to her, to wear upon her own fair person. Next my hands and fingers were mumbled [*sic*], and declared to be as soft as a child's, and my hair was likened to a lion's mane.[35]

What is notable is the importance of touch in the encounter. Ungugu is interested not just in how Speke's body looks but in how his skin, hair, and clothes feel. Speke's cooperation with her curiosity is remarkable as well. The woman's royal status perhaps made such a close examination more acceptable than it would have been in other contexts.[36] But his acquiescence was also a reflection of power relations: Speke complied because he needed the ruler's permission to proceed, a fact that he emphasized in his published account.[37] In doing so he tried to present himself as motivated by geographical exploration and portrayed the fact that he was willing to be examined as exemplary of his determination to reach his goal. In another unpublished telling of the incident, Speke sought to reassert control over the situation through dehumanizing and denigrating the woman. He claimed that she was "a short stumpy, and pugnosed dame, of considerable age, almost in her dotage," and that her hands were "more like bear paws than human digits."[38] When retelling

FIG. 28. "Our Camp outside the Palace," from Speke's *Journal of the Discovery of the Source of the Nile*. One person can be seen peering into the hut. Courtesy of New York Public Library. Image ref: 1258399.

the incident, Speke could use such rhetorical moves to regain his position as observer and evaluator. But while traveling he clearly depended on cultivating local rulers' interest in his own body. The intimate and multisensory ways Ungugu examined him illustrate her power over him more viscerally than a focus on vision and gazes.

Speke's cultivation of curiosity in his own body is even more apparent on his later expedition with Grant (fig. 29). During their stay at the court of Mutesa, the kabaka of Buganda, Speke seems to have encouraged the king's interest in his body. These occurrences have particular significance when viewed in the context of the broader changes happening in East Africa in this period. Recent scholarship on the history of Buganda has devoted attention to the country's engagement with external forces in the second half of the nineteenth century. By the time Mutesa took the throne in 1857 or 1858, Buganda had achieved "unprecedented size and influence" within the lake regions.[39] At the same time, though, Muslim traders from Zanzibar were starting to travel to the country in growing numbers, bringing with them guns and cloth. These traders were, therefore, "a new and potentially destabilising element in local and regional politics," threatening Ganda supremacy.[40] Throughout his reign Mutesa sought to use contact with both Swahili traders and European missionaries and travelers (of whom Speke was the first) to secure his position domestically and advance the power of the Ganda state in the region. He often used such encounters to try and obtain arms, ammunition, and other militarily useful technology. Viewed in this context, these encounters are representative of a broader process of unequal engagement with the outside world and the king's attempts to use their encounters to his own advantage.[41]

Mutesa's motivations were as complex as those of Speke. The explorers' entertainment value was part of the reason for his hospitality. Mutesa was particularly interested in Speke's hair, clothes, and skin. Speke reports that at one of their meetings he gave Mutesa "a tortoise-shell comb to comb his hair straight with, as he invariably remarked on the beautiful manner in which I dressed my hair, making me uncap to show it to his women." At another meeting, Speke reported, "my hair must now be shown and admired, then my shoes taken off and inspected, and my trousers tucked up to show that I am white all over."[42] These incidents foreground Speke's total dependence on his hosts and his willingness to allow examinations of his body to maintain good relations. In his unpublished account, Speke hinted that such displays of his body were a regular occurrence: "The brothers' wives then wish to see me and come before us, when [*sic*] I have to take off my hat and shoes as usual, my ready compliance induced the princes to say, 'we like Bana [Speke's nickname in the court], very much: he is so pretty, gentle and, good-natured.'"[43] Speke's insistence that Mutesa and others found his body attractive should be viewed with some skepticism. But these passages do illustrate how often and intently his body and dress were examined and the important role this played in securing the continued hospitality of his hosts.

Similar forces were at play in quite different ways during Mutesa's first examinations of Grant's body. His hands attracted the king's attention because he had lost the thumb and index finger on his right hand while serving in India.[44] Mutesa asked how Grant had received the injury, a curiosity that Grant suspected was a product of the fact that amputation was a common punishment for criminals in Buganda. Mutesa's concern

FIG. 29. James William Grant. Stipple engraving by S. Hollyer after Urquart. Courtesy of the Wellcome Collection. Image ref: 37571.

shows that travelers' bodies were implicated in debates over trustworthiness and credibility in the field. It also demonstrates that Mutesa was far from naive in his interactions with the foreigners. Like British commentators he examined the explorers' bodies to assess if he could trust what they said. But after Mutesa had established Grant was not a threat, he expected him to display his body in ways like Speke. Grant wrote: "He [Mutesa] signalled that I was to show my head uncovered to the ladies; a titter followed, and all of us laughed heartedly. Another signal, and I was told to place my hat on."[45] Here we see the degree of control Mutesa exercised over them during their stay at the court. Displaying their bodies on command was clearly something the explorers were expected to do in return for the welcome they received. Their ready compliance with such requests is suggestive of their relative lack of power.

Mutesa also used Speke's presence to actively bolster his own domestic political authority. During their time together, the king asked Speke for items of European clothing so "that he may appear like Bana."[46] Speke had some trousers made for Mutesa and gave him some of his own clothes, including a waistcoat.[47] On several occasions Speke and Grant encountered the kabaka wearing these outfits.[48] Speke did not think such dress suited the king. More interesting than Speke's reaction is the fact that Mutesa thought these clothes were "very becoming" and showed them off to his relatives.[49] These displays suggest that he adopted new forms of dress not to impress Speke but his court and subjects. These incidents are particularly noteworthy because the kabaka saw himself as a "corporeal reflection of the country's body politic."[50] It is worth remembering that Speke (dishonestly) represented himself as a member of the British royal family, and, based on this misinformation, Mutesa may have thought his clothes were items of royal dress and, therefore, appropriate for his status.

Focusing on Speke and Grant at the Ganda court demonstrates that they were the objects of examination by African rulers, who both saw explorers as a source of an entertainment and as an opportunity to acquire knowledge and resources. In some cases displaying his own body was the price Speke had to pay for access to local leaders. Thus these incidents draw attention to African agency and how European exploration was shaped by economic and political changes taking place in nineteenth-century East Africa. The examination of Speke's body in the field should not blind us to the powerful and negative effects of Burton's and Speke's published writings about Africans, but they do highlight how the power of such observations was in their circulation within the British Empire. In the field bodily examination was a two-way process.

Similar trends also emerged in Isabella Bird's travels through Japan in 1878 and China in 1894–97. Here, too, bodily examinations were shaped by the political and cultural changes taking place in the regions she visited. As she traveled throughout Japan, Bird often found herself examined by large crowds, sometimes up to one thousand strong.[51] The curiosity shown toward Bird was in part a product of the wider interest in European, and particularly English, models of development at the time.[52] After the Meiji restoration of 1868, Japan went through a period of dramatic economic, political, and social upheaval. These changes involved the modernization and Westernization of the Japanese economy, education system, and government and saw growing numbers of Europeans travel to Japan.[53] Japanese interest in Bird was not only motivated by the fact that she looked, dressed, and spoke differently; she was also a representative of a society of growing interest to Japanese people at the time.[54]

In private Bird portrayed such incidents in fairly neutral terms. In a letter to John Murray written while traveling, Bird suggested that the "only annoyance" of Japan was "the continuous mobbing by polite crowds."[55] She claimed that they were generally "quiet and gentle."[56] But she reported occasional problems with more aggressive mobbing that left her unable to breathe.[57] These incidents were also featured in her book.[58] The examinations reverse many of the power dynamics at play in the rest of Bird's work. She found herself with "no privacy," stared at, and observed by large crowds of people who were as interested in her as she was in them.[59] In public Bird drew on racial stereotypes when writing about these incidents, describing crowds as "filthy and squalid."[60] In this way she portrayed Japanese people as an unpleasant and deindividualized mass rather than as autonomous human subjects. Such language was an effort to reassert control and counter the loss of power she experienced in the field. Bird also made a series of ethnographic observations, particularly of Indigenous Ainu people, during her time in Japan. These helped to ensure she was positioned as the observer rather than the observed in her published account.[61]

Bird generally showed little interest in the motivations of those who looked at her. But her account does include one curious incident that suggests Japanese people viewed her body as strange (and possibly nonhuman) when she traveled in more remote areas. She reported encountering a group of Japanese boys who "said they thought Ito [her guide] was a monkey-player, i.e., the keeper of a monkey theatre, I a big ape, and the poles of my bed the scaffolding of the stage!"[62] Bird included these comments to demonstrate the unreasonable and unscientific nature of Japanese curiosity, presenting the boy's ideas as an amusing quirk borne of ignorance. But the comments do suggest that many Japanese people looked at Bird's body in ways that threw into question her assumptions about racial hierarchy. The comments also convey the impression that even these children were aware of her dependence on her guide. The potential for alternative and troubling readings is perhaps the reason the children's comments were removed from the revised, single-volume, popular edition of her book, which condensed her account in ways that made it more fully conform to tropes of travel writing.[63]

The examinations of Bird's body were not confined to distanced stares. Her descriptions of the curiosity of her guide, interpreter, and companion Ito Tsurukichi allow us to see much more nuanced forms of encounter and the importance of sound as well as vision in such incidents. Bird reported that he used the trip with her as a learning opportunity, keeping an extensive diary in both Japanese and English, recording observations about the areas through which he traveled. He also used his time with Bird to develop his English vocabulary, avoiding any expressions that she considered either "common" or "slangy."[64] Tsurukichi's actions could be read as a form of mimicry, as examined by Homi Bhabha.[65] Yet, while he was keen to improve his English vocabulary, he was not in awe of white bodies or British habits. Bird wrote that "our manners, eyes, and modes of eating appear simply odious to him."[66] Such comments suggest that while he was keen to improve his English vocabulary and translation skills, he had little interest in mimicking other aspects of British culture. His response might be better understood as mockery.[67] But there are also reasons to think his learning was done with a specific purpose.

Tsurukichi put his cultural knowledge and linguistic skills to use after Bird left Japan. After Bird's departure, Tsurukichi became a prominent and successful interpreter and guide in Japan working for other European visitors. When he died in 1913 at the age of fifty-seven, he was the subject of numerous obituaries in Japanese papers, which referred to him as "the pioneer of guiding in Japan and Japan's leading interpreter" and noted that he had "struck up familiarity with many gentlemen of repute." Some prominent Western travel company executives were "so taken with his abilities and intelligence that they allowed him the privilege of travelling free on their trains and ships."[68] Tsurukichi's success as a guide and translator also raised broader questions about who was leading who throughout the journey (as discussed in chapter 3). For Tsurukichi studying Bird allowed him to develop his abilities as a translator and guide—useful skills, given the growing European involvement in Japan in the period. Broader political trends also shaped how Bird's body was examined when she traveled to China in the 1890s. This time, though, Bird was inspected in much more aggressive ways.

VIOLENCE AND HOSTILITY

While explorers often complained about invasive interest in their body, it could often be very useful. They generally operated beyond the formal boundaries of European empires, so they found it more difficult to travel when they encountered the hostility of local populations. In situations where explorers had more lethal weapons, larger parties of armed men might crush such resistance through violence (a tactic particularly common among European explorers of Australia).[69] But this was not a viable strategy when explorers traveled in a smaller group or in regions with powerful local states. In such areas hostility could halt the progress of an expedition.

These issues are particularly apparent in Bird's travels in Korea and Japan in the mid-1890s. There she experienced resistance to her presence, including shouts, violence, and other invasions of her privacy. As in Japan local people often saw Bird as a representative of Britain and Europe, more broadly. But being a foreigner and a Christian had much more negative connotations, given the broader political context in the country and growing resistance to Western exploitation of China in the second half of the nineteenth century. Partly this opposition was a backlash against the growing influence of Christian missions, which divided communities and were a "direct challenge" to the power of local elites as well as Chinese religion and ethics. More broadly, it was a product of "the economic hardship and political humiliation" that imperialist powers inflicted on China in the nineteenth century.[70] This involved a series of treaties (imposed through military force or threats of it) that handed over the administration of many coastal ports to European powers, fragmented the Chinese state, allowed the import of opium, and gave foreigners extraterritorial privileges—a situation that caused famines and economic privation.[71] Bird's presence in the country was the product of broader imperial power relations, particularly the extraterritorial privileges Europeans had while traveling and living in China. These developments led to growing anti-foreign sentiments and the emergence of Chinese nationalism. Tensions erupted soon after Bird left China, in the form of the Boxer Uprising (1899–1901), a rebellion that sought to free the country from foreign encroachment and that involved attacks on both Chinese Christians and Western missionaries.[72]

Bird's writings show how Chinese resistance to European domination limited her ability to travel and observe. Bird directly reported the anti-colonial and anti-foreign resistance she encountered in her book *The Yangtze Valley and Beyond* (1899). She claimed that "the literary class" were often "brutal and insulting, and generally tended to excite hostility against the foreigner."[73] She also described being abused in ways that reflected the growing anti-Western sentiment in China at the time. Private correspondence with her editor added further detail to this story. She wrote that she was pelted with stones and called a "foreign devil," a "foreign dog," and a "child eater." When attacked in Liang-shan, Bird claimed, she would "have been beaten to death had it not been for the timely appearance of a mandarin with soldiers."[74] This is a revealing admission, as it shows that, at times, facilitating her passage through China relied on the explicit threat of state violence.

In such incidents patriarchy and anticolonial resistance interacted in complex ways. Traveling in an open chair (as Bird did) was very unusual for women in China, meaning that Bird's "audacious appearance combines social and political insult, as it . . . suggests China is weak enough to be presided over or conquered by a woman."[75] Her attempts to use dress to blend in were not altogether successful. She was not prepared to fully adopt Chinese dress (such as wearing trousers) and combined items of Manchu clothing with "English tan shoes" and "a Japanese *kurumaya*'s hat" and "English gloves."[76] This strange combination inevitably attracted attention: many Chinese people viewed her costume as "a confusion of foreign inferiority."[77] Even Bird's hagiographic biographer, Anna Stoddart, conceded that her choice of chair and hat "was perhaps rather indiscreet, as her appearance roused a certain suspicion."[78] Bird's identity as a British woman, the way she traveled, and the political context meant that local people often greeted her with sticks, stones, shouts, and insults. Rather than a countergaze, she met a combination of anticolonial and patriarchal violence.

Chinese resistance shaped where Bird traveled, limiting her ability to see certain areas. She often avoided cities, as they were particular hotbeds of anti-foreign sentiment. Even in larger towns she reported that "a woman-traveller must shut herself up rigidly in her room from arrival until departure unless she desire to provoke a row."[79] When she arrived she often found that crowds of people assembled "shouting, ofttimes hooting, and fighting each other for a look at the foreigner." Men and women occasionally tried to enter her rooms, but she wrote that a more common problem was "the slow scraping of holes in the plaster partition" between rooms, "accompanied by the application of a succession of eyes to the hole, more whispering, and some giggling."[80]

In her published account, Bird sought to reassert some control over the situation by complaining that Chinese "curiosity appeared singularly unintelligent" and fantasized about poking a revolver or syringe through the hole.[81] In the field, though, she became wearied and frustrated by such encounters. At one point she met a mandarin who invited her to stay at his home. She turned down the invitation, claiming "grave doubts what my reception might really be in the women's quarters, and I dreaded the stifling curiosity succeeded by the stagnation of dullness."[82] This incident shows how the "uninhibited expression of curiosity extended to Bird is so irritating that she forfeits an unparalleled opportunity to view Chinese home life."[83]

Despite these measures Bird still found her body examined in ways that she found uncomfortable. She reported that at one inn in Szechuan, the landlord's wife, and "often a bevy of women," would "come and pick over my things." They initially asked Bird questions about her life and motivations for travel, "but they soon tired of the trouble of interrogating me" and talked to Bird's interpreter instead. Bird wrote that their conversations with him often focused on her "'ugly eyes" and "big feet" and the fact that they thought her hair was "like wool."[84] They were apparently interested in the fact that she could write, as they had not seen a literate woman before.

At another inn Bird reported, she had to "gratify the curiosity" of nearly thirty women who had come to see and examine her. "My hair, 'big feet,' shoes and gloves were all a great amusement to them," she wrote.[85] The women also commented on the coarse texture of her hair, suggesting that they may well have touched her, as happened elsewhere. In *Korea and Her Neighbours* (1898), she described similar examinations in more detail, reporting that, there, women and children also invaded her room. Once inside they "examined my clothing, took out my hairpins, and pulled down my hair, took off my slippers, drew my sleeves up to the elbow and pinched my arms to see if they were of the same flesh and blood as their own; they investigated my few possessions minutely, trying on my hat and gloves."[86] What Bird found challenging about such forms of examination was not that just that she was being observed, but that these observations saw her privacy and personal space invaded in ways she found "aggressive and intolerable."[87] Being gazed at was one thing, but having your possessions poked, your hair touched, and your room entered was quite another.

Within these encounters the line between observer and observed was far from clear cut. Previous scholarship has demonstrated that Bird struggled in China, because she encountered a reciprocal Chinese gaze that got in the way of her "desire to observe freely."[88] In contrast I suggest that using the metaphor of the gaze in this context obscures the actual ways Chinese opposition to her travels was expressed. In Japan Bird had been stared at, a level of curiosity she did not find very troubling. In China she was pelted with stones and shouted at, and her possessions (and possibly her body) were poked and prodded. Chinese resistance was expressed through more than just a gaze.

The metaphor of "the gaze" is perhaps most helpful in reflecting on the ways that, in writing about her experiences, Bird retrospectively asserted her authority. She continually disparaged the motivations behind Chinese curiosity, and she used the hostility she experienced to demonstrate her own vulnerability and bravery in ways reminiscent of the "anti-conquest" trope discussed by Pratt.[89] Describing incidents of violence allowed Bird to draw attention away from broader imperialist involvement in China, portraying Europeans as victims rather than exploiters. Yet a close reading of the diverse ways Bird's body was examined offers insights into how many Chinese people viewed her body—not as a heroic explorer, but as a strange and barbaric embodiment of European imperialism. Bird's experiences in China demonstrate how explorers could come unstuck when they encountered directly anticolonial sentiments and overt resistance to their presence. Given the hostility she experienced, it is unsurprising that Bird traveled across China with a revolver concealed under her clothing. She never used it and commented that "it is most unwise to let fire-arms be seen in Chinese travelling."[90]

Not all travelers hid their weapons. When they thought they could win, British explorers sought to overcome the resistance of local people through force. Explorers in Australia had long used such tactics to overcome Aboriginal resistance to settler encroachment. In the last decades of the nineteenth century as the balance of power shifted in favor of Europeans, these tactics were also applied on a growing number of African expeditions.[91]

In August 1875 Stanley carried out a premeditated attack on Bumbire Island in Lake Victoria. The first time he had visited, Stanley complained about the hostile examinations he experienced at the hands of the islanders: "One of them even gave a tug of my hair, thinking it a wig."[92] Following this relations continued to sour, leading to a violent confrontation in which Stanley claimed to have killed between nine and fourteen people.[93] When Stanley next sought to pass through the area, he attacked the island without warning, killing forty-two islanders.[94]

Stanley's act of "cold-blooded revenge" provoked fierce debate in Britain. One writer denounced his attack on the island as "exploration by warfare" while another suggested that his journey had involved "exploration plus buccaneering."[95] Much of the controversy centered on the question of whether violent tactics could still be considered an act of exploration in the tradition of other "heroic" travelers. The *Saturday Review* was unconvinced: "The memory of [Mungo] Park and [David] Livingstone, and of the discoveries which they made, the toils which they endured, reproaches this new mode and these motives of assault on unknown countries."[96] While offering a rather idealized image of previous expeditions, ignoring acts of violence by Livingstone, the *Saturday Review*'s comments illustrate the relatively narrow parameters in which "exploration" could be carried out. To many vulnerability played a key role in defining scientific travel from other forms of engagement, such as warfare and more direct colonial intervention. These remarks bring to the surface the implicit reliance of many less contentious travelers on the acquiescence, and curiosity, of local rulers and populations. As Adrian Wisnicki has suggested, we might read many of the "blank spaces" on European maps as expressions of African agency and resistance.[97]

Whereas Burton and Speke had sought to travel through these areas through negotiating passage with local rulers, by the 1880s and 1890s growing numbers of explorers resorted to more direct forms of violence. They thus played a role in the expansion of European empires into the interior of the continent in this period.[98] Stanley's Emin Pasha Relief Expedition (1886–89) kidnapped and ransomed African women and children in areas where the local population refused to trade with them.[99] But, as African expeditions became increasingly piratical, it became much harder to draw any distinction between exploration and unprovoked conquest. Unpopulated Antarctica provided a space in which the *Saturday Review*'s mythologized image of exploration as peaceful scientific quest could be performed.[100] Even there explorers still found their bodies poked, prodded, and measured. But these examinations shed light on a different set of power relations.

EXPLORERS' BODIES AS OBJECTS OF SCIENTIFIC STUDY

Bodily examinations were a common feature of polar expeditions. Since the eighteenth century at least, naval medics and commanders had used long sea voyages to investigate the causes of scurvy, experimenting with different diets.[101] In the nineteenth century, the experiences of polar explorers were used to study the effects of diet upon the body.[102] Regular examinations were also an important way of checking on the health of a ship's crew. On both of Robert F. Scott's expeditions to Antarctica, medics examined the men's bodies in the hope that the observations would shed light on the physiological impact of the polar climate. Here explorers were far more willing to submit to scrutiny by white scientists than they were by Asian or African people. But such practices were not unidirectional: many working-class sailors seem to have viewed the inspection of their bodies as a source of amusement.[103]

Practices of scientific examination are particularly visible on the British National Antarctic Expedition (BNAE, 1901–4). During the Antarctic winters, the expedition's surgeon, Reginald Koettlitz, carried out routine examinations of the explorers' bodies, assisted by the expedition's junior surgeon, Edward Wilson. These inspections, they thought, might provide some broader insight into how human bodies performed in Antarctica and the impacts of the environment.[104] Scott described their comprehensive nature: "Our weight is taken, and then measurements of chest, filled and empty, waist, calf, forearm and upper arm; then, by means of a small spring instrument, our power of with right and left hand is recorded, and finally the capacity of our lungs is measured as we discharge one long breath into the spirometer."[105] The surgeon also took blood samples from both the officers and crew, using a pin to prick the men's little fingers.[106] He then studied these samples, counting the number of red and white cells.[107]

Although the composition of Antarctic expeditions was comparatively homogenous, such practices were still shaped by broader power relations. The examination of officers and men took place at different times and in different spaces. The officers and scientific staff were examined in the wardroom; the crew were inspected the following evening on the ship's mess deck.[108] These separate procedures were a product of naval disciplinary norms and class-ridden distinctions between officers and men.[109] Officers' accounts suggest that they witnessed the measurement of the crew's bodies, meaning that they could scrutinize the sailors' bodies in a nonreciprocal way.[110] The lack of reciprocity is symptomatic of a broader medical gaze that sought to monitor and control the bodies of the sailors.[111] But it is important not to lose sight of the material dimensions of these examinations that saw different bodies brought into close proximity. Later Koettlitz and Wilson were called on to examine the explorers' bodies for signs of scurvy, inspecting their bodies visually but also touching them. In October 1902 Koettlitz reported that one of the cooks might have the disease because he had a bruise on his leg which had "a decidedly suggestive feel."[112]

The biography of the BNAE's surgeon throws into question the idea that we should understand these examinations as only medical in nature. Before joining the BNAE, Koettlitz was the "doctor, geologist, anthropologist and field manager" on Herbert Weld Blundell's expedition to East Africa (1898–99), which traveled through Somaliland, Abyssinia, and Sudan.[113] On the expedition Koettlitz took anthropological observations of the Galla people. He measured various aspects of their bodies, recording their average chest, arm, and head measurements as well as their height. Koettlitz presented his anthropological findings upon his return and published these findings in the *Journal of the Anthropological Institute* and the *Journal of the Royal Geographical Society*.[114]

The similarities between Koettlitz's measurement of bodies on this expedition and in Antarctica were not lost on members of the BNAE. A cartoon (fig. 30) and satirical article appeared in the expedition's newspaper, the *Polar Times*, in which the monthly inspections were referred to as "anthropological measurements."[115] In the image the difference between the surgeon's thin, gangly body and the muscular and masculine sailor is emphasized. The cartoon highlights the degree to which the surgeon's interest in the crew's bodies was considered strange. By satirizing Koettlitz's tall, thin body, the drawing turns him into an object of amusement and suggests that his examination of other bodies was linked to the physical inadequacy of his own. Albert Armitage, the expedition's second-in-command, claimed that the examinations "always created a lot of merriment" and joked that Koettlitz "had an unquenchable desire for our blood."[116] Koettlitz's work in this area, therefore, led to his body and methods becoming the object of humor, much to his frustration.[117] The crew's response to these examinations shows that, as elsewhere, examinations were not a one-way process, but were met with a reciprocal gaze and satire.

The results of these measurements proved inconclusive, showing "no sign of any change in general condition."[118] The failure of the expedition to produce any definitive findings did not end efforts to use Antarctic exploration as a way to study the body. Before Scott returned to Antarctica in 1910, all members of the expedition had their skin, hair, and eye color recorded, in order to establish if "men of special colouring were more adventurous than their fellows."[119] The measurements were taken on a tintometer, an instrument developed by Joseph William Lovibond to systematically measure colors. The device has a history of its own within anthropology and was used by a University of Cambridge expedition to the Torres Straits Islands (1898) in order to establish if color perception was universal.[120]

The idea that an individual's skin and hair tone could have an impact on their performance in a cold climate is illustrative of the idea of "race" within medical knowledge in this period; it is also suggestive of a similarity in methods between anthropology and medicine. But there is an interesting crossover of methods between animals and humans. Scott's thinking on this issue may also have been shaped by his readings about Ernest Shackleton's ponies. On his *Nimrod* Expedition (1907–9), Shackleton had ponies of various colors with him, several of which died before the southern journey. Shackleton reported the "rather curious fact that the survivors were the white or light-colored animals, while disaster had befallen all the dark animals."[121] Drawing on these ideas Scott took only light-colored ponies on his expedition, a fact that may have contributed

FIG. 30. "Anthropological Measurements," a cartoon from the *South Polar Times* vol 2, June 1903, 46–48. Reproduced by kind permission of the Syndics of Cambridge University Library.

to the inferior quality of the animals purchased.[122] As these debates illustrate, physical appearance played a significant role within Edwardian explorers' thinking about extreme environment survival, suggestive of the influence of scientific racism in extreme environment physiology in this period. These tests are also interesting in that they saw the methods of anthropology used on the bodies of white men.

The Lovibond test was not the only form of experimentation on the human body carried out during the expedition. As Vanessa Heggie has examined, the explorers later conducted "an extended self-experiment on diet and metabolism" during a winter sledge journey to collect emperor penguin eggs.[123] The expeditions thus employed a strange mix of both anthropological and physiological measurements to understand the explorers' bodies. Because of the very different context, the explorers' reaction to these examinations differed from the response of other travelers to African and Asian curiosity. In Antarctica the explorers willingly submitted to examination by white scientists. Even so, examinations were not unidirectional and could be contested through humor.

CONCLUSION

European expedition leaders were not the only ones producing new knowledge. Previous writing has shown that travelers often met with a countervailing reciprocal gaze. This chapter has developed these arguments, showing that such encounters involved whole bodies and multiple senses. Explorers found their bodies, poked, prodded, and touched by local people, often for complex reasons that reflected local social and political circumstances. We also see the limitations of analyzing all forms of examination using broad metaphors like "the gaze." In many instances explorers could cope with being looked at by others, but they struggled to maintain their patience when they found their bodies examined more closely. In East Africa, for instance, Speke found stares amusing but objected much more vigorously when such examinations crossed a certain line, reacting with violence. In Japan Bird sometimes grew frustrated by the large crowds she encountered but found that they did not disrupt her journey. In China she was the object of curiosity, violence, and heckling, which severely limited her ability to travel freely. In analyzing such incidents together, I highlight the need not only for more nuanced discussions about the different forms of bodily encounter but also for an investigation of cross-cutting themes.

Focusing on these practices of examination in more detail draws attention to the agency of African, Asian, and Indigenous people. Due to the limited and partial nature of the expeditionary archive, explorers' accounts require a close and critical reading. But it is clear that their ability to travel was shaped by historical, social, and cultural understandings of their bodies in the areas they traveled through and by broader political developments. While my reading of such incidents is suggestive rather than definitive, it nevertheless demonstrates that explorers were not the only ones producing knowledge—a fact that serves to blur the lines separating the explorer and the explored.

The response of explorers to these practices of examination varied considerably, depending on the relations of power at play in each incident. Speke complied with and even encouraged curiosity in his body, which foregrounds his dependence on local rulers while traveling. Explorers in areas where this curiosity was absent encountered far more problems. Bird was mostly powerless when faced with Chinese resistance. In contrast Stanley's use of force to kill and intimidate local populations defending their land had a lasting impact on his reputation and signals the increasingly blurry line between exploration and invasion. Explorers adopted different strategies when it came to local interest in them, but it was always something that they had to contend with.

In Antarctica different power relations came into play. The explorers were more willing to submit to examination by white scientists. Even so the officers would not allow their bodies to be examined in front of the men, and Koettlitz practices of examination were satirized. Allowing your body to be examined put explorers in the uncomfortable position of the observed, and even in the comparatively homogenous setting of Antarctica this was something that seemed strange and amusing—leading to questions about the motivations of the examiner.

Focusing on such incidents offers insights into the relationship between examination, power, and writing. In their published works, explorers frequently complained about and disparaged the motivations behind the African and Asian interest in their bodies. Writing became a way for them to reassert control over what had happened. It often seems that the more vulnerable they felt in the field, the more they tried to emphasize their authority by drawing on offensive or objectifying language about those who examined them. In many ways this chapter brings attention back to many of the issues discussed in chapter 1. Both at home and in the field, explorers found their bodies examined in multiple and unpredictable ways. They could try and keep these two spaces separate, writing about their experiences in ways acceptable to domestic audiences, but the realities of travel still left indelible marks on their bodies and their narratives.

Writing too freely about reciprocal examinations could cause problems. In *The Journal of the Discovery of the Source of the Nile* (1863),

Speke described an incident where he examined a woman's body in the Kingdom of Karagwe. The woman was "unable to stand except on all fours" as she was very large, for, according to Speke, "fattening is the first duty of a fashionable female life" in this area.[124] Speke reported that he was "desirous to get a good view of her, and actually to measure her, and induced her to give me facilities for doing so, by offering in return to show her a bit of my naked legs and arms."[125] In a passage in an unpublished draft of the book he described the reaction to this display: "Of course my arms and legs had called forth immense raptures of delight and exclamations of beauty, for no one would believe before that the whole of my body was white."[126] As such comments show, Speke cultivated curiosity in his body, using it as a tool to encourage reciprocal examination. However, his editors thought some of his descriptions too troubling to include in his published account. Reciprocity might be acceptable in the field, but it caused problems on the page. Even with the account sanitized, the armchair geographer James MacQueen still complained that Speke "seems to delight and excel in such exhibitions" of his own body and used the incident to attack his wider credibility.[127] The problem was not just that Speke's body had been examined, but that he seemed to have enjoyed it. Domestic critics were all too aware that expeditions might also provide opportunities for sexual encounters, the topic to which I now turn.

7

SEXUAL BODIES

In July 1861 John Hanning Speke penned a report to the secretary of the Royal Geographical Society (RGS) from East Africa. The letter described his faltering progress toward Lake Victoria—a body of water he was desperate to prove as the source of the Nile. Speke outlined the problems his expedition had experienced so far and the lessons he had drawn from them about successful travel. He stated that, if he were to return to East Africa, he would not employ an Arab headman of his party. Speke claimed that the man who had performed this role, Sheikh Said bin Salim, was a "nuisance," as "he requires luxuries and women."[1] In his subsequently published book, Speke suggested that bin Salim was a less effective traveler because he missed the "beautiful wives" he had left behind.[2] In such passages Speke sought to show how differently he approached travel from the Arab merchants who had preceded European explorers in their journeys into Central Africa. Unlike them, a scientific explorer, Speke suggested, had no need for women or romance. "A rolling stone gathers no moss," he elsewhere explained.[3] Arab traders, he implied, were different because they became embroiled in sexual relationships on their journeys and traveled with their partners. This distinction was, for the most part, a question of reputation. As I show below, Speke had numerous relationships with African women. His comments reveal that the idea of exploration rested on the scientific traveler being an unattached "man of action" driven forth by the search for scientific and geographical truths—always maintaining some physical and emotional separation from local people.[4]

This chapter shows that, in reality, many masculine explorers in Africa engaged in sexual practices considered deviant at home. These relationships generally had abusive dimensions but could involve physical and emotional intimacy. Expeditions could also be the site of sexual regulation, with explorers seeking to control the bodies of unmarried women who joined their parties. Women explorers and men on polar expeditions had their sexualities regulated in complex ways as well. In this sense explorers' sexual bodies were shaped by the different gender expectations and social circumstances they confronted.

Studying sex on expeditions illustrates many of the fundamental challenges of explorers' writings as historical sources. The amount of information about such matters varies considerably, with some writers discussing sex much more than others. Problematic elements were often removed or recast in more favorable terms to render narratives legally publishable and morally acceptable to domestic readers, but such redaction was uneven. Even censored accounts give us important insights into what went on, and a reading of private sources can show us elements that explorers sought to hide. The image that emerges is different from the idealized, heroic bodies that I began this book with. Instead explorers appear to us in a variety of guises, varying between lusty, loving, vulnerable, abusive, and violent.

Historians and geographers have devoted significant attention to the role that sex (and its regulation) played within European empires.[5] Much of this scholarship has been concerned with the efforts of the British government and reformist campaigners to regulate sex and the sex trade in both Britain and colonized countries.[6] The exact contours of these efforts varied considerably in different contexts. But colonial governments made growing efforts to regulate and suppress interracial relationships over the course of the nineteenth and early twentieth centuries. This reflected a changing understanding of the concept of "race," which in the eighteenth century represented "a much more diffuse set of differences" than it did by the mid-nineteenth.[7] The growing British prejudice against interracial relationships—and broader Christian norms that privileged heterosexuality, marriage, and monogamy—meant that explorers had good reason to avoid writing about their sexual encounters on expeditions. As we shall see, men who gave too much away often suffered reputational damage, though in some quarters there seems to have been a tacit toleration of such activities, provided they were kept private. The picture was different when it came to white women (particularly those from an upper- and middle-class background), who faced far greater constraints: their social standing could be destroyed by any hint of extramarital or interracial sex.[8]

While there were good reasons not to write about sex, the mass appeal of many travel narratives rested on their descriptions of non-European sexual practices. As Edward W. Said has argued, these forces were so powerful that travelers' writings are often full of relatively "regulated and uniform" tropes. Explorers in the Middle East often focused on "harems, princesses, princes, slaves, veils, dancing girls and boys, sherbets, ointments, and so on." Many European writers depicted the Orient as a place where they could escape domestic constraints and experience "the freedom of licentious sex." There was an element of truth in the idea that travel offered European men opportunities to avoid increasingly restrictive domestic codes. But accounts of "Oriental sex" also became a standardized and marketable commodity.[9]

Similar—though importantly different—tropes permeate writing about Africa in this period. There is a long tradition of European writers portraying Africa as a site of sexual excess, focusing particularly on African women. Anne McClintock terms this vision of Africa as the "porno-tropics" and argues that it is a frequent theme in European depictions of the continent and its people.[10] As elsewhere, focusing on explorers' private writings can help to give a more balanced picture. But such sources are not always available, as many travelers destroyed private records that showed them engaging in taboo sexual activities.[11]

Another challenge is that some explorers tried to use discussions of sex to shock domestic audiences in ways that make it hard to uncover the truth. Richard Burton wrote about sex more than any other explorer and consequently features prominently in scholarship of sexuality and empire.[12] Perhaps because of his own enthusiasm for courting controversy, it is often hard to separate fact from fiction in Burton's writings. One striking example is an incident described in the "terminal essay" of his translation of the *One Thousand and One Nights* (1886). In the essay Burton reported that, while a solider in Sindh in the mid-1840s, there was a brothel near his camp "in which not women but boys and eunuchs, the former demanding nearly a double price, lay for hire." At his commanders' orders, he had visited the site in disguise "and obtained the fullest details, which were duly dispatched to Government House."[13] He claimed that the report had fallen into the wrong hands and had a negative impact on his career. Burton included lurid details that may have been gained through personal participation in sexual activities. But no copy of the report has ever been found, and Burton's biographers are divided over the factual accuracy of this anecdote, and the suggestion that it affected his professional reputation.[14]

While it is hard to establish the veracity of the story, the context in which this tale was reported tells us something more interesting: it appears in a broader essay on anal sex, in which Burton argued that male homosexuality was more prevalent in warm climates. In doing so he discussed the issue in scientific rather than moral terms, thus using an analysis of sexual practices in the Orient to shift public discussions about sex and sexuality in Britain. In later life Burton translated several texts dealing with erotic themes, including the *Kama Sutra* and *The Perfumed Garden*, publishing some of his most contentious materials privately, under a pseudonym. He did this because Parliament's Obscene Publications Act of 1857 outlawed the publication of pornography and other works considered morally corrupting. But, as the exact parameters of what was banned were not made explicit, it also stifled the publication of numerous texts dealing with sex and sexuality, as publishers did not wish to risk prosecution. Burton viewed his own interest in such matters as a "radical" act, pushing the boundaries of intellectual inquiry.[15]

Burton's writings often went beyond what was considered legally and morally publishable. Passages dealing with sex were edited out of *Personal Narrative of a Pilgrimage to Mecca and al-Medina* (1855) prior to publication by the Egyptologist John Gardiner Wilkinson, who described them as "unpleasant garbage."[16] Similar issues emerged when it came to the publication of *First Footsteps in East Africa* (1856), an account of his expedition through the Horn of Africa to the city of Harar (now in Ethiopia). Burton penned an appendix to the work, which discussed details of female genital mutilation practiced in the area. Again the description given raised the possibility that he had gathered these details firsthand. Although the text was written in Latin to limit its readership, Burton's publishers thought the account was unsuitable for publication and ordered that it be "ripped out."[17] The appendix was only ever included in a few, apparently private, copies of the book. Censorship shaped what was advisable and legal to publish about sex on expeditions.

Despite the efforts of his editors, Burton's published travel accounts contain numerous (and frequently offensive) comments on the inspection and examination of women's bodies. On his 1853 trip to Arabia, Burton sometimes disguised himself as a Muslim medic, giving him particular opportunities for examining the bodies of those he encountered. Burton wrote about one incident during his stay in Cairo, where he treated a group of Abyssinian "slave girls" who were suffering from "consumption, dysentery, and varicose veins." He described their bodies in terms that drew on sexualized and dehumanizing racial stereotypes, claiming the women were "average specimens of the . . . Abyssinian breed, broad-shouldered thin-flanked, fine limbed, and with haunches of prodigious size." Burton then discussed the women's manner of "flirtation" with him.[18] In such passages Burton portrayed the Orient as a sensual and sexual place, reflecting broader conventions of travel writing about the region.[19]

These comments were far from isolated. In *The Lake Regions of Central Africa* (1860), Burton described an incident where he encountered a group of Nyamwezi women. He claimed that the women's "dress—a short kilt of calabash fiber—rather set off than concealed their charms, and, though destitute of petticoat or crinoline, they were wholly unconscious of indecorum. . . . These 'beautiful domestic animals,'" Burton went on, "graciously smiled when, in my best Kinyamwezi, I did my devoir to the sex; and the present of a little tobacco always secured for a me a seat in the undress circle."[20] Such descriptions illustrate his portrayal of African women in objectifying and dehumanizing terms. They also show that while traveling Burton felt free to have sexual experiences that would have been impossible at home.

While risqué—and, to most modern readers, offensive—such comments continue to reinforce the idea of Burton as an emotionally detached, if sexually curious, observer. Although he wrote publicly about the bodies of African people, he did not publish anything about his sexual relationship with his wife, Isabel Burton (née Arundell).[21] Biographers' assessments of their intimate life are based only on the most fleeting passages.[22] He wrote about his marital relationship in terms very different from that he adopted with racialized "others." In this sense Burton's writings differed from those of other travelers in the detail given to sexual matters, but they also reflect broader conventions in important ways.

Focusing on Richard's travels with his wife allows us to see expeditions as sites of emotional intimacy and perhaps even sexual experimentation. Richard and Isabel maintained a close relationship throughout their marriage, collaborating on numerous literary projects. They also traveled together in Syria, Palestine, and South America, and Isabel penned and published her own accounts of such trips.[23] At the time these journeys were generally understood as "travel" rather than exploration—in part because, as we have seen, exploration was often portrayed as a specifically masculine activity. These expeditions saw the couple try on new forms of dress and try out different gender roles. When they traveled together in the Syrian desert, Isabel donned men's clothing, posing as Richard's son. "It *sounds* indecent," she commented, "but all Arab clothes are so baggy and draping that it little matters whether you are dressed as a man or woman."[24] It is difficult to establish if the motivations for this role-playing were sexual or if they reflected the broader Victorian fascination with role-playing and impersonation.[25] Either way, travel allowed the couple to experiment with crossdressing, and they clearly seem to have enjoyed this.

Isabel and Richard were not the only couple who traveled together. David and Mary Livingstone undertook even more extreme journeys, but this did not always prove straightforward. During their work as missionaries in Southern Africa, they traveled together. They even brought their infant children on an 1850 expedition that aimed to contact Sebituane, the Kololo chief who ruled an area north of Lake Ngami. In this case Mary Livingstone was five months pregnant at the start of what they anticipated would be an arduous journey involving two crossings of the Kalahari Desert. The party were without water for days and afflicted by diseases, including a bronchial infection. Mary gave birth shortly after the party's return, but the baby soon died. Mary's parents thought that David's decision to take his pregnant wife on this journey had been ill considered. Mary's mother wrote to Livingstone criticizing his judgment and cautioning him against bringing his family on future expeditions in the strongest possible terms.[26] David and Mary ignored this plea and made a second attempt in 1851, this time with more success.

These journeys were not Mary's last involvement in exploration. Over the following years, she spent much of her time in Scotland caring for the couple's children. She planned to accompany David on the Zambesi Expedition (1858–64) but discovered she was pregnant on the outward journey (suggesting that she and David had been recently intimate). She left the expedition at Cape Town to give birth to the couple's sixth child and eventually returned to Scotland. Unhappy there, Mary had an "intense" desire to join her husband who was still in East Africa.[27] She again traveled out to join the Zambesi Expedition, reaching the river in January 1862. By this time Mary Livingstone had become a heavy drinker, partially in response to the hardships and loneliness she had suffered while her husband was away.[28] Drinking seems to have weakened Mary's immune system, and she died from fever on April 27, only three months after joining the party. Deeply affected by her death, David "wept" extensively and reported that the loss "lies like a heavy weight on my heart."[29] Reflecting on Mary Livingstone's involvement in exploration reminds us that expeditions could be the site of physical and emotional intimacy, and that pregnant women sometimes took part in them. It also illustrates some of the problems explorers faced in trying to maintain relationships while spending long periods overseas.

Sometimes couples traveled together under circumstances that were less domestically acceptable. In 1861 Florence and Samuel Baker left Cairo leading an expedition that eventually reached Lake Albert, claiming it as a source of the Nile. The wealthy Baker was in many respects a typical British explorer of this period, but Florence's backstory was complex: born Barbara Maria Szász in Transylvania, she was orphaned in the violence of the 1848–49 Hungarian Revolution. She then fled to the Ottoman Empire, where she was enslaved before being sent for training in a harem and renamed Florenz. At around fifteen years old, she was sold in a slave market in the Ottoman city of Vidin (now in Bulgaria). Samuel tried to buy her, but was outbid.[30] He then bribed a guard to facilitate Florence's "escape," and the couple were soon informally married. Through modern eyes the decision of Baker (who was almost forty at the time) to marry a recently enslaved teenager seems like a case of exploitation rather than emancipation, given the tremendous disparity of power between the two of them. Even in Victorian England it was highly irregular. Few of Baker's class would "think of marrying a girl before she came out in society, at seventeen."[31] There was also the question of why Samuel had been at the slave market in the first place.

Samuel's relationship with Florence was a product of the inequality of power between them. As we shall see, this was far from an aberration. More unusual was that, following the expedition, Samuel and Florence traveled back to Britain together and married formally (for an image of Florence in Britain, see fig. 31). After leading a further expedition (this time sponsored by Khedive Ismail of Egypt), the couple lived out much of their married life in Devon. Lady Baker was well provided for after Samuel's death in 1893.[32] Despite their efforts to regularize the relationship (and to hide the most dubious aspects of the circumstances under which

FIG. 31. Lady Baker as photographed by Henry Maul ca.1875. Wikimedia Commons.

they had met), gossip would dog the couple for the rest of lives. Due to patriarchal cultural norms, Florence was more affected than Samuel. In 1867, after Samuel had become a famed and knighted explorer, Queen Victoria refused to allow Florence to be presented at court because of "rumours of intimacy before marriage."[33]

While it caused issues in high society, in other circles Samuel's relationship to Florence only enhanced the popular appeal of his journey. Florence appeared alongside Baker when he received the RGS's Gold Medal and featured prominently in his book on the journey.[34] Her presence differentiated their expedition from those made by male travelers, and her unique backstory contributed to the popular appeal of the account. As one contemporary review noted, "the constant presence of Mrs. Baker in the story" made Samuel's book "more marvellous" and gave the narrative a "chivalrous interest."[35]

Another review remarked that "Baker himself . . . is not so interesting as his wife."[36] But some commentators questioned the advisability of taking a woman on an expedition, illustrating efforts to keep exploration as a solely masculine preserve.[37] Such patronizing attitudes belie the important role that Florence played in the journey because of her ability to speak Arabic, which she had learned while enslaved. This proved invaluable as they traveled through Egypt and Sudan, and she used it both during day-to-day leadership of the expeditionary party and in negotiations with local rulers.[38] She also spent time teaching the language to Samuel.[39] It is hard to overstate the important role she played in facilitating the couple's journey. The reliance of explorers on women's skills was not confined to Africa. Many arctic explorers learned important survival techniques and local languages through exploitative sexual relationships with Inuit women.[40]

SEX, ENCOUNTER, AND EXPLOITATION

Explorers sometimes engaged in domestically taboo sexual practices.[41] In the field they were often far away from other Europeans and had significant power over the women and children they encountered, which they sometimes used to sexually exploit them. They seldom wrote about such issues in public, but their private and unpublished records tell a different story.

Speke's unpublished first draft of his *Journal of the Discovery of the Source of the River Nile* (1863) contains a level of detail missing from most other explorers' accounts and from his own subsequently published (and heavily edited) book. In the manuscript, Speke recorded that he became sexually and emotionally involved with an African teenager, failing to maintain any degree of detachment. During his time at the Bugandan court, Speke was "given" two enslaved Wahuma "girls"—whom he named Kahala and Meri—as a present by the queen mother. It is fairly clear that Speke had sexual relations with the older of these two girls.[42] In this manuscript Speke recalls being informed by the queen mother that the younger one, Kahala, was about twelve and "too young for present purposes." The elder, Meri, was "about eighteen years or so" and "in the prime of youth and beauty, very large of limb [and] dark in colour."

Meri was clearly not pleased about being forced into a relationship with Speke and was "sobbing bitterly" as she walked back to his hut.[43] In this sense the encounter was a product of Speke's alliance with the Bugandan royal family, and the comparative lack of power of the enslaved teenagers. Speke's private descriptions of their time together strongly imply that Speke and Meri had a sexual relationship; it is unclear if this was consensual. At one point Speke claimed that Meri at first had a "maidenly modesty," but he also reported that he spent one night "taming the silent shrew," raising the possibility that Speke raped her.[44]

Speke and Meri lived together for about a month. Speke described himself during this time as a "henpecked husband," and his writings indicate his infatuation with Meri. His feelings for her were made most explicit when the couple were "divorced." She became sick, and the relationship broke down. She claimed that she needed a goat sacrificed to restore her health. Speke thought that she was inducing sickness by putting her fingers down her throat—an action that gives some inkling of the distress that this situation was causing her. Speke reacted strangely, trying to have the M'ganga (medic) who had prescribed this treatment flogged, which further angered Meri. After this fallout he claimed that she "showed neither love nor attachment for me," implying that she had displayed some before the incident. The argument left Speke remorseful, and he reported that the turmoil left him with "an overflowing" and subsequently "nearly broken" heart. He attempted to win her affections back with a shower of gifts, but to no avail. The situation, Speke wrote, "nearly drove my judgement from me . . . which broke my sleep and destroyed my rest."[45]

The incident is revealing in that we see Speke in such different terms from the passages with which I started this chapter. Far from being an independent "rolling stone," Speke became deeply attached to Meri, behaving as if driven by an unhealthy and jealous infatuation. It was this passion that, Dane Kennedy argues, was the most troubling aspect of the incident for domestic audiences: "To succumb to the power of the heart in such a context was to betray the social values and ideological commitments the explorer was expected to embody."[46] But we should also consider the impact of this episode on Meri. Even through Speke's one-sided account, we can see something of her anguish at being forced into a relationship with a much older stranger.

Speke's disturbing relationship with Meri was probably not his only sexual encounter with African women. In another passage that was heavily toned down in his published account, he reported that when helping one of the Bugandan king Mutesa's wives (named Lubuga) across a river, he was distracted by her "imploring face and naked breasts" and "voluptuous captivating manner." Speke speculated that there was at least some mutual curiosity, and that Lubuga was "anxious to feel . . . what the white man is like." We will never know if Speke's assessment of her feelings was accurate. Other incidents do seem to show that Speke was an object of curiosity while in the Ganda court. Speke flirted with another woman, named Kariana, whom he described as a "very pretty little woman" who followed him and began to stare at him. He thought—somewhat big-headedly—that she was "spellbound" with his "beauty." Speke described how a "flirtation" then ensued in which she proceeded to examine his "hair, watch, the contents of my pockets—everything." At Speke's direction the couple then walked along, arm in arm, "as if we had been in Hyde Park rather than in central Africa, flirting and coquetting all the way." Near Speke's hut Kariana was told to return to her home, but she left "begging" that Speke would come and see her again.[47] Such passages illustrate a more two-directional romantic curiosity than we find in Speke's relationship with Meri. There is, of course, a need to treat Speke's one-sided (and somewhat self-aggrandizing) report of the encounter with a pinch of salt. Yet we do see that he clearly enjoyed the opportunities that travel offered for such flirtatious intercultural encounters of a less abusive nature.

During his time at the Ganda court, Speke appears to have been viewed as an expert on sexual and relationship difficulties. Speke was called on by King Mutesa to "examine his water-works" as the monarch was concerned about his sexual potency and ability to have children.[48] Speke found nothing wrong with him and sketched his naked body (fig. 32).[49] He gave him detailed and specific advice on sexual performance and conception, instructing him to sleep with his wives shortly after they had their period, "as the seed vessels were more sensitive then, and to refrain from overindulgences, which destroyed the appetite in early youth and soon rendered a fine man impotent."[50] He also cautioned him against the overuse of stimulants (probably alcohol), suggesting that they would "rather tend to diminish his natural powers."[51] Mutesa, however, had doubts about

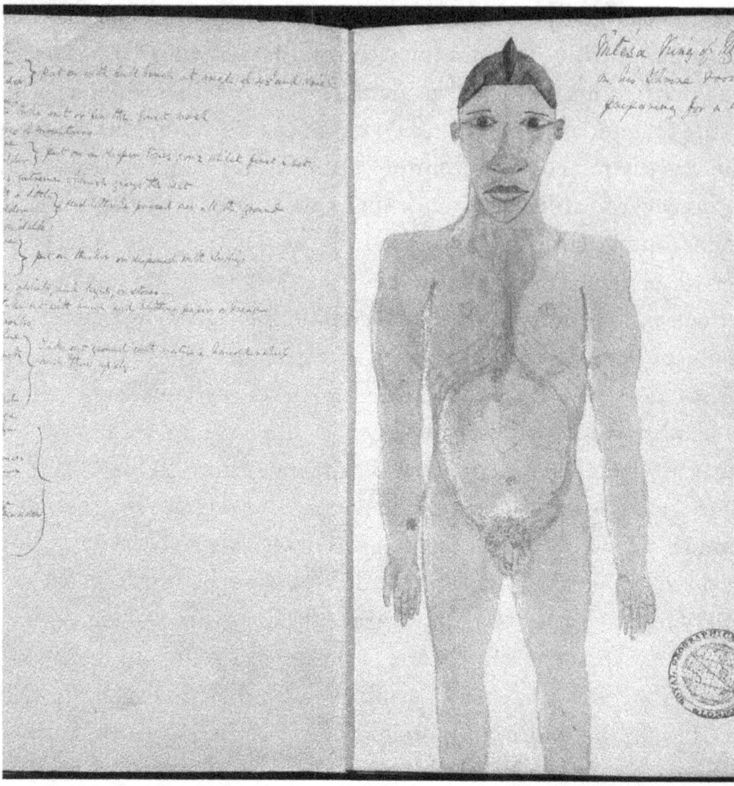

FIG. 32. Mutesa, King of Buganda, in his throne room preparing for a blister. Reproduced with permission of the RGS-IBG.

this ability to sexually satisfy women and "could not believe in a short stick being so good as a long stick, because the longer one could reach so much farther, whilst the short one would only knock about the doorway." Within a few days, Speke was called upon to help treat men suffering with "the inferior venereal disease."[52] At first glance these passages might seem to be easily incorporated into a standard colonial narrative about African sexual promiscuity and indiscipline. But, read from a different angle, they show that Speke possessed quite a detailed knowledge about venereal diseases and could instruct the king on how to sexually satisfy and impregnate his wives. One also wonders what it was about Speke's conduct that made Mutesa turn to him for advice in the first place.

These various incidents were heavily edited in the final published version of the book, due to the work of Speke's publisher, William Blackwood, and John Burton, an editor employed by him.[53] Both men removed problematic passages and rephrased incidents in language more acceptable to domestic audiences. But even the censored version was too much for Speke's critics, particularly the armchair geographer James MacQueen, who attacked Speke by implying that he had sexual relationships with African women. MacQueen argued that Speke's conduct had been inappropriate, and that he had always sought "the youngest and the most handsome" female porters.[54] He included such information to raise questions about the explorers' character and, in doing so, cast doubt on the credibility of his geographical and anthropological claims.[55] We can make sense of this example through the framework of book history, which allows us to understand how travelers' raw accounts were edited into marketable commodities through the labor of multiple people.[56]

But writing was not the only way that news about Speke's encounters spread. Henry Morton Stanley later reported that Livingstone had told him about an encounter in which Speke was questioned by the Duchess of Sutherland about the nudity of some African women. Speke described their state of undress in such a vivid manner that he apparently "caused several ladies to cry, 'Oh! Captain Speke!' And the Duchess to hide her face with a fan."[57] Given the duchess's curiosity about the matter, her modesty at Speke's response might have been exaggerated. Forming sexual relationships with African women and girls might help an explorer cope with the isolation of the field, but such encounters were considered impolite (if possibly titillating) conversation at home.

The lasting damage done to Speke's reputation by these rumors of intimacy gives us a better understanding of why explorers seldom wrote about interracial sex. But his relationship with Meri also shows us the uneven terrain of power on which such interactions were conducted. Did other explorers also abuse their power and engage in acts of sexual violence? Often it is impossible to say. Occasionally, we have clearer evidence gleaned from travelers' private journals. The actions of the officers of the rear column of the Emin Pasha Relief Expedition (EPRE), who traveled up the Congo in 1886–89, are recorded in a level of detail absent elsewhere. Running short on supplies, the rear column camped at Yambuya, while the advance column, led by Stanley, pressed on to Equatoria to try and rescue the beleaguered Ottoman governor, Emin Pasha. When waiting for Stanley to return, the European officers of the rear column kept a group of Manyema women in a state of sexual slavery, referring to them as their "cannibal concubines."[58] In March 1888 William Bonny, who penned the rear column's journal, reported that he and two other officers purchased three women off Arab traders—one in exchange for a "pair of long boots."[59] In one misogynistic passage, Bonny complained about the higher price initially demanded, claiming that he had paid less in England "to call a woman mine." A few days later, Bonny purchased a different woman—named Subitaki in his account. She was, he estimated, between eighteen and twenty years old and could speak "a few words" of Swahili.[60]

Even from the one-sided account available, we can see that the women were unhappy with their situation and wished to return home to their villages. On Sunday, March 11, they planned a coordinated escape. Two of the women were able to run away and were reported to still be on the run several days later; the third woman was recaptured. Bonny wrote that he saw another officer pointing a revolver at her, threatening to shoot her if she tried to escape again.[61] Her fate was not recorded. But Bonny did note that a further three enslaved women—whom the rear column's commander, Edmund Musgrave Barttelot, claimed to own—"ran away," suggesting that the officers purchased several groups of enslaved women.[62] The actions of the officers of the rear column of the EPRE seem extreme even by the standards of Victorian explorers. But it is equally difficult to know if other explorers engaged in similar practices but avoided writing such details down.[63]

When he found out about their conduct, Stanley sought to distance himself from the officers and criticized their general conduct. Yet he also ensured that their sexual crimes were never made public, avoiding any mention of them in his book or speeches.[64] Perhaps Stanley was silent because he may have engaged in related activities himself. On his expeditions he "proved to be a serial collector of African boys"; he also took a number of African children to England. Although we will never know the precise nature of such relationships, "sexual predation" was a likely motivation.[65]

Even if they did not involve sexual abuse, these relationships with African adolescents were highly exploitative. On his Livingstone Search Expedition (1871–72), Stanley liberated an enslaved boy, originally called N'dugu M'hali, and renamed him Kalulu, subsequently employing him

FIG. 33. Stanley with Kalulu and Selim, depicted in an 1872 wood engraving. Courtesy of the Wellcome Collection. Image ref: 561189i.

as a gun-bearer.[66] Stanley took Kalulu to England and used him to publicize his books and lectures (fig. 33).[67] Kalulu then returned to Africa as part of the explorers' next African journey, eventually drowning in a canoeing accident on the River Congo.[68] The unfair nature of Stanley's relationships with teenage boys was noted by his contemporaries. Burton commented that Stanley's Palestinian translator on the same expedition—a teenager named Selim—was paid poorly and "had to travel the whole way on foot."[69] Later Stanley tried to use two African children to promote the Stanley and Africa Exhibition in Regent Street, London, in 1890. The child's guardians were twice taken to court by another member of the exhibition committee and the Anti-Slavery Society, who claimed that the boys had been "cruelly treated and would be returned to slavery in Africa." The second of these actions was successful, and the two boys had new guardians appointed before being sent to an "industrial school in South Africa."[70] Even by Victorian standards, Stanley's treatment of these boys was shocking.

It is difficult to prove definitively that Stanley sexually abused any of these boys. But his fiction gives more insight on his sexual desires. In 1873 Stanley published a novel, *My Kalulu, Prince, King, and Slave*, which tells the story of the relationship between two adolescents, Kalulu and Selim (both very loosely based on the real people noted above). In the novel Selim is a fifteen-year-old Zanzibari Arab who joins his father's trading caravan into the interior. Kalulu, also a teenager, is in line for the throne of an African kingdom. Selim is enslaved on his journey into the interior before escaping and being rescued and nursed back to health by the prince. The pair develop a close relationship and eventually form a "ceremony of brotherhood."[71] Kalulu is then enslaved and liberated in Zanzibar by Selim. Writing a fictional narrative about African people allowed Stanley "to voice desires indirectly, at a greater remove than could be achieved in the first-person travelogue."[72] There are also strong homoerotic themes in the book. Stanley described Selim and Kalulu's bodies in terms that "linger on their physical beauty" and the boys' relationship seems to go beyond the normal bounds of friendship.[73] Drawing on this book and passages in his autobiography, several authors have argued that Stanley had homosexual desires. Though, in doing so, such scholars (particularly the biographer Frank McLynn) have sometimes relied on fragmentary evidence and "contested" psychoanalytic theories.[74] I am also uncomfortable that such writers have sought to link Stanley's apparent sexual interest in (and possible abuse of) African juveniles with homosexuality, as this feels uncomfortably similar to homophobic discourses that have sought to liken same-sex attraction to pedophilia.[75]

REGULATED SEXUALITIES

Expeditions were also environments of sexual regulation for both explorers and those they led. Single women who tried to join African expeditions were the target of particular attention. During his last expedition, a woman named Ntaoéka chose to join Livingstone's party while he was searching for the source of the Nile in Central Africa.[76] "I did not like to have a fine-looking woman among us unattached," he wrote, so he encouraged her to wed one of the unmarried men in his party. Livingstone was perhaps keen to see her married to keep his own desires in check and to avoid gossip concerning his relationships with African women, which had emerged previously. He was clearly attracted to Ntaoéka; he noted in his diary that she "has a good presence and is buxom."[77] After an earlier expedition, one newspaper quoted several of his descriptions of women he had met and noted that "he seems to have a keen eye for the beauty of

East African ladies."[78] Upon finding Livingstone in East Africa, Stanley felt the need to publicly correct the "facetious gentleman's story" that the lost explorer had married an African princess.[79] Livingstone may have also been worried about the conduct of other men in his party and the potential for jealousies to emerge.

Either way Ntaoéka was initially unwilling to marry. "Circumstances led to the other women wishing Ntaoéka married," Livingstone reported, "and on my speaking to her again she consented."[80] He does not record exactly how many other women were with the party and what "circumstances" led them to wish this. But his private diaries suggest there was particular animosity between her and Halima, his cook, perhaps suggesting romantic rivalry over the affections for Halima's husband Amoda.[81] On June 3, 1872, after further persuasion, Ntaoéka married James Chuma, a young man of Yao heritage who had been liberated from slavery by Livingstone some ten years earlier and who had previously worked on the Zambesi Expedition.[82] The incident illustrates how explorers and African expedition members saw unattached women as a problem within an expeditionary party; it also shows the important role explorers played in the management of intimate relationships and jealousies. That Livingstone encouraged Ntaoéka to marry one of his closest traveling companions suggests that controlling women's bodies could be a means to encourage monogamy and loyalty among his male followers.

These kinds of relationships were far from uncommon. On expeditions in East Africa, Nyamwezi porters and Waungwana askari also formed and maintained relationships with women. Some were based on slavery or kidnapping. We get a sense of the various kinds of relationships in an expeditionary party by focusing on the partners of Speke's "faithfuls" when they arrived in Cairo in 1863.[83] The party included women and girls who were photographed in the Egyptian capital by Louis Royer, who had established a studio there (fig. 34).[84] Their names were Sikujua, Kahala, Mzizi, and Faida. Faida was later married to Frij, one of the expedition's cooks (suggesting that she was unmarried when the photo was taken).[85] James Augustus Grant reported that Sikujua was "young, but very black" and joined the expedition after being captured by one of Speke's men in Buganda: "No one ever came to claim her, and she remained his property. She had the pretty oval face and large ears of the Wahuma." She did not participate in the expedition, out of choice. Kahala was the younger

FIG. 34. Sikujua, Kahala, Mzizi, and Faida as photographed at Cairo depicted in Speke's *The Journal of the Discovery of the Source of the Nile*. Courtesy of the New York Public Library. Image ref: 1258429.

enslaved girl given to Speke by the queen mother of Buganda (discussed earlier in the chapter).[86] She may well have worked as a translator, as Speke noted that she was being taught Swahili in order to act as a "medium of communication" between Speke and Meri.[87] In an unpublished draft of his book, Speke reported that, toward the end of his time in Buganda, he gave Kahala, who was then about twelve years old, "over to [Sidi Mubarak] Bombay as wife; for she prefers playing with dirty little children to behaving like a young lady and has caught the itch."[88] This quote gives some sense of her age and immaturity at the time. Grant also wrote that she stayed with the expedition as "the property of Bombay"—language that frames her marriage to Bombay as a question of ownership.[89] Kahala does not seem to have been happy with this arrangement, as she tried to run away and return to Speke's hut after she was forced to live with Bombay.[90] Once out of Buganda, Kahala would have had more reason to stay with the party, because, if she left the expedition, she might well have been enslaved by whomever she sought shelter with.

The women and girls who traveled all the way to Cairo clearly had limited autonomy, given their age and the circumstances under which they joined the party. But such individuals are not necessarily representative. Many free women from Unyamwezi were involved in the East African caravan system, joining expeditions and trading parties by choice. These temporary "caravan marriages" were an established part of East African travel cultures. Unlike those photographed at Cairo, free women often had a significant amount of agency, as many Nyamwezi women were able to limit their fertility and achieve "some negotiated advantage by attaching themselves to higher status males."[91] As discussed in chapter 2, such women performed a variety of important roles, carrying goods, building shelters, and cooking food. Explorers' accounts often include fleeting references to such relationships. In *The Journal of the Discovery of the Source of the Nile*, Speke reported that "many" men in his party "married" on their journey through East Africa.[92] As a nod to domestic sensibilities, he claimed that he tried to prevent such relationships. Due to the small number of women who arrived in Egypt, we must assume that many of these relationships were temporary. Expeditions were not solitary, masculine endeavors but "mobile communities."[93]

As in any community, romantic relationships can be a source of tension. Expedition leaders frequently became embroiled in managing such relationships, sometimes seeking to impose Christian monogamy. We find a telling example of this in Joseph Thomson's *To the Central African Lakes and Back* (1881). Near Lake Nyasa (now Lake Malawi), the expedition's headman, James Chuma, discovered that there was an "unattached female in the camp, who . . . had clandestinely joined our caravan with the wonderful aspiration of reaching the coast." Thomson reported his dilemma: "What was he to do with her? This seemed to be rather a difficult problem to solve. If we cast her out of the camp, she would inevitably be made a slave of and yet it would be demoralizing to have such a woman with us." Drawing on Livingstone's experiences (possibly relayed to him by Chuma), Thomson decided that the best course of action would be to marry the woman to a member of his party. He thought this might prove difficult, however, as he thought the woman unattractive. The men were assembled and Chuma explained the situation, stressing the woman's physical strength:

> There had been a general burst of laughter at first. But at last Stambol [a porter], who had always shown himself very fond of domestic comforts, was overpowered by the argument that a pair of good hands were in the long-run better than a pretty face, and much less liable to be run away with. So he boldly offered to marry her. This settled all difficulties at once, and in less than half-an-hour we married them right off. Thereafter followed the customary festivities. It may be added that Mrs. Stamboul stuck faithfully to the caravan till on our return march to the coast, when lagging somewhat behind one day she disappeared, and was no more heard of.[94]

The passage is revealing, in that it shows the degree to which white expedition leaders and culturally hybrid headmen such as Chuma felt entitled to regulate the bodies of African women and girls. That said, the fact that the woman in question seems to have escaped when they approached the coast illustrates how these efforts were not totally successful. Explorers' control was never total.

So far I have spoken mostly of the carnal encounters of masculine explorers. How different was the situation for British women? As we have seen in the case of Florence Baker, women's sexual interactions were subject to far greater scrutiny and censure than those of men. Sex was only acceptable within wedlock, but marriage could make it hard for a woman to travel independently. On her wedding day, a woman lost a litany of financial and legal rights, making it challenging to travel without a husband's presence or permission. Many women explorers avoided marriage to protect their independence.[95] The explorer and botanist Marianne North commented that marriage was a "terrible experiment" that reduced women to the status of "a sort of upper servant to be scolded if the pickles are not right."[96]

Isabella Bird did not marry until middle age and turned down numerous proposals. After the death of her sister, Henrietta Bird (with whom she was extremely close), she married John Bishop, the medic who had cared for her ailing sibling. As elsewhere we know almost nothing about the sexual dimensions of their relationship, but we do know that she took steps to ensure that she retained the ability to travel. Her choice of husband seems aimed at maintaining her autonomy and power: Bishop was of lower social standing than Bird, who had distant familial connections to aristocracy.[97] Before they married she forged an "understanding" with Bishop that, if she needed a change, she would still be able to carry out further "outlandish travelling."[98] A man would not have had to get such assurances.

Women were also expected to adopt different standards of behavior in the field. Guidebooks aimed at upper- and middle-class women, such as *Hints to Lady Travellers* (1889), suggested that sexual harassment could be avoided through dressing modestly and behaving femininely.[99] Imperial culture also placed far greater constraints on interracial sex between white women and Black men, and, in their writings, many women travelers explicitly spoke out against such relationships. Their authority in the field came from stressing their racial rather than their gender identity, and interracial marriage would involve "control by the very same men over whom they had so enjoyed asserting their newfound authority."[100] Given these broader cultural constraints, it is unsurprising that women travelers wrote about intimate relationships in far more restrained terms.

Bird's writings frame her desires in romantic rather than sexual terms. In *A Lady's Life in the Rocky Mountains* (1879), Bird included a description of her liaison with a man called Jim Nugent, a notorious Colorado trapper and settler. During her time in the mountains, Bird and Jim traveled together on horseback and had long and deep conversations. Her attraction to him is clear even in her carefully composed published account. She described him in flattering terms, noting that while one side of his face bore the marks of a bear attack, he still had "large grey-blue eyes, deeply set, with well-marked eyebrows, a handsome aquiline nose, and a very handsome mouth."[101]

Bird's private correspondence illustrates that their relationship was more than fraternal. But, even in private, we see this framed more in terms of love than sex. In November 1873 Jim confessed to Bird that he had become "attached" to her and "it was killing him. . . . I was terrified. It made me shake all over. He is a man whom any woman might love" she commented, "but no sane woman would marry."[102] It is impossible to know if there was any sexual dimension to their relationship. Because of Bird's strong religious faith, it seems highly unlikely. But, if there was, she would have good reasons not to write about it.

The relationship, while probably sexless, did involve emotional intimacy, and Bird clearly had strong feelings for Nugent, who died soon after she left America.[103] Bird later reported that she had been visited by his ghost while in a hotel room in Switzerland at almost the same time that he died.[104] Bird also used the story to add romantic interest to her books. As one contemporary review noted, she portrayed him as a "romantic ruffian" and "never tired of dwelling" on his "genius" and "chivalry."[105] But she was careful not to go too far. She claimed that she had encouraged Nugent to abandon his life of drink and sin and seek salvation in Christ. In this sense the narrative is ultimately a morality tale in which Bird was able to successfully discipline her own desire and even tries to reform the behaviors of Nugent.

If they carefully navigated Victorian gender regulations, women travelers could benefit from including some hints of romance in their accounts. But even these suggestions of attraction could be too much for some readers. One reviewer, writing in *John Bull*, while impressed with Bird's achievements, desired that other women should not follow in her footsteps.[106] The love interest in Bird's account perhaps also marked the book as an account of "travel" in comparison to the larger and more scientific works that she published in the 1880s and 1890s.[107] In her later writings, which were aimed more at securing her reputation as a scientific observer (and thereby at a more masculine readership), Bird avoided discussing any similar romantic interest. In this sense we see that, in order to become a scientific explorer, Bird had to cast herself in quite different terms from the romantically and emotionally engaged (if sexually chaste) traveler of her early works.

We find similar challenges in writing about Antarctic exploration. These expeditions were masculine environments, and the first woman did not set foot on the continent until 1935. In a society in which homosexuality was illegal, keeping Antarctica a place free of women meant that it was generally portrayed as a space free from sex or romance. The traveler and biographer Sarah Wheeler suggests that, for Robert F. Scott, the "Antarctic possessed a virginity . . . that provided an alternative to the spoiled and messy world" outside it.[108] The desire to preserve Antarctica as a womanless—and ostensibly sexless—place seems to have been shared by Ernest Shackleton. A group of schoolgirls applied to join his Imperial Trans-Antarctic Expedition (1914–17), and a group of nurses asked to join the Shackleton-Rowett Expedition (1921–22). In the latter case, the nurses professed to be "young, very strong" and unafraid of any hardship.

They implored Shackleton to give them a "sporting chance & don't rule us out just because we are women."[109] Neither application was successful.

We should not accept at face value the idea of Antarctica as a space free of sex and desire. Although Shackleton prevented women from taking part in Antarctic expeditions, he spent at least some of his time in Antarctica thinking about them. On his 1907–9 attempt to reach the South Pole, Shackleton probably named a number of newly discovered features after women with whom he had close and potentially romantic relationships.[110]

It is also possible that some Antarctic explorers engaged in sexual activity or relationships with other men. But there is no direct evidence. Suggestions that some polar explorers had same-sex desires have sometimes been motivated by homophobia rather than efforts to write a more inclusive history. As Max Jones has demonstrated, the 1985 television series *The Last Place on Earth* sought to cast Scott as a bad leader and blunderer. But, in doing so, the show drew on homophobic ideas and "alluded to the homosexuality of the British elite and hinted at Scott's 'Nancy Boy' qualities."[111] The program tells us much more about the cultural and sexual politics of 1980s Britain than it does about attitudes around sex on Edwardian expeditions.

There is little direct evidence of homosexuality on Edwardian Antarctic expeditions. We might start to understand why this is the case by viewing the journeys in the context of the Royal Navy, the institution from which many polar explorers were drawn. The navy had traditionally adopted a stricter approach to punishing homosexual acts than civilian authorities. Court martial records show that men were punished for engaging in prohibited sexual activities. But convictions were quite rare, as the burden of proof required at a court martial was seldom met.[112] Often men were investigated when they were found to have anal syphilis. This did not prove, of course, that any sexual encounter had been consensual, so men suspected of such acts were routinely dismissed from the service through administrative measures rather than by trial at court martial.[113]

As I discussed in chapter 3, Antarctic explorers went through a comparatively rigorous selection process compared to other travelers that involved physical examinations and the taking up of references (which sometimes discussed syphilis infections and, implicitly, men's sexual conduct). This process would clearly have discriminated against any men suspected of same-sex desires, making it far less likely that they would have been able to join an expedition. Once in Antarctica men would also have found it hard to hide any sexual activities, as most expeditions overwintered in small and relatively cramped huts in which there was little privacy; the case of the British National Antarctic Expedition (BNAE) is somewhat different, as the explorers overwintered on board the *Discovery*. What we see, therefore, is that Antarctic explorers' sexual bodies were far more regulated than those of travelers in other contexts. The fact that we find little evidence of same-sex intimacy on these expeditions is not accidental.

We confront similar issues when it comes to masturbation on expeditions. For quite understandable reasons, explorers' accounts seldom mention this practice. But is there any evidence that explorers engaged in it? One possible hint can be found in Shackleton's *The Heart of the Antarctic* (1909). In it he reports that, in the expedition's hut, the explorers all had a curtain in front of their sleeping bunk in case any individual wished to "sport their oak."[114] Shackleton's biographer Ranulph Fiennes argues that this phrase was a public-school euphemism for masturbation.[115] But, given that the term appears in Shackleton's published narrative, it is unlikely that this is his intended meaning. He probably used it in its more conventional sense—that is, a sign to other members of the expedition that the inhabitant of the bunk wanted some privacy.

The ways that other explorers wrote about masturbation also makes it unlikely that Shackleton would so freely hint at this practice. About a decade before Shackleton's Antarctic expedition, Frederick Jackson led the Jackson-Harmsworth Expedition (1894–97) to the Arctic Archipelago of Franz Josef Land. While his writings offer far more clues that explorers masturbated, they also show how this practice was associated with immorality. In his unpublished diary, Jackson reported problems with a young man called John W. Heyward, who carried out duties such as cooking and cleaning in the expedition's hut.[116] Jackson and Heyward had never seen eye to eye. But toward the end of their time in the Arctic Jackson described how he and the expedition's doctor, Reginald Koettlitz (who later served on the BNAE), talked about Heyward, "who in the opinion of both of us is practicing a horrible habit upon himself which accounts for his extraordinary and seemingly lunatic ways." Koettlitz informed Jackson that he had confronted the man about this practice "some time ago and Heyward 'blushed crimson' and did not deny it." Writing in his private diary, Jackson continued:

> I consider that I should be to blame, as the youth will become worse and worse it will effectively ruin him if he does not put an end to it, if I do not do my best to stop it. The doctor tells me that moral suasion is the only remedy that can be applied; so we agreed that we should talk to him, pointing out what a horribly contemptible practice it is and which dogs only would indulge in, and the extremely serious results of it. This he did this evening. Heyward did not deny it to him even and was on the point of blabbering over it when he lectured him. What an infernal nuisance men are![117]

Although this passage is somewhat euphemistic, it is likely from the content, context, and the man's reported reaction that this unfortunate man had been discovered masturbating. What is most striking about Jackson's diary entry is the degree to which both he and Koettlitz appear disturbed by the practice, viewing it as exemplary of a general lack of moral fiber, with potentially "serious results" for his mental health. If explorers did masturbate, it was absolutely essential that they kept it secret. As Jackson's stern moralizing terms illustrate, self-discipline was a seemingly widespread cultural expectation. But the fact that Heyward was caught also shows the difficulty of keeping secrets in the confined space of a polar hut. Unsurprisingly, this passage never made it into Jackson's published accounts of the expedition. Nor is there any mention of the fact that the captain of the expedition's ship suffered from syphilis, causing him to be sent back from the Arctic after the first winter.[118]

On polar ships explorers' sexual bodies were clearly far more regulated than those of travelers in other contexts. Here the field is not a site of freedom but a location of enhanced surveillance. That said, we can take this line of argument too far. Explorers' published accounts show them experimenting with cross-dressing and engaging in physical horseplay. Such incidents suggest that the isolated masculine environment provided opportunities for them to engage in behaviors they avoided at home. "There is no doubt that the sailors dearly love to make up," Scott summarized in *The Voyage of the* Discovery.[119] On Antarctic expeditions this often started with the "crossing the line" ceremony, a maritime initiation rite for those crossing the equator for the first time. The specifics of the display varied somewhat from expedition to expedition, but as a general rule the ceremonies involved fancy dress, often including some cross-dressing. For instance, on the *Terra Nova*, Petty Officer Browning was dressed as the Greek goddess of the sea, Amphitrite.[120] Cross-dressing continued when the explorers reached Antarctica, often forming an important part of midwinter celebrations. Some explorers seem to have enjoyed these festivities tremendously, while others felt uncomfortable, suggesting that some saw the potentially transgressive side of such actions.[121]

Incidents of fun and camaraderie were part of a broader effort to create a "family of men" and a domesticated, homosocial, and homely environment on an expedition.[122] But practices of fancy dress also had more troubling aspects, which involved explorers wearing blackface (fig. 35). On the BNAE, the expedition's first lieutenant, Charles Royds, organized a "minstrel show" with twelve men from the mess deck.[123] The men wore sewn-together suits "of the most vivid colours and grotesque forms" and manufactured wigs. In their performance, they blacked up and spoke in crude impersonations of African American accents.[124] The shows clearly illustrate how Antarctic exploration was influenced by the racist ideologies of empire.[125] The literary critic Tomasz Filip Mossakowski claims that the jokes of the show also "drew attention to the lower stratum of the body, namely, the belly, butt, and sexual organs, known for their reproductive functions and excretions." He even suggests that the shows saw the white explorers try on the "black penis."[126] While there may be some truth in these ideas, this argument risks overstating the sexual dimensions of these performances. Many jokes focused not on sex but on the quality of food the sailors received and the work they were asked to carry out.[127] Sex on Antarctic expeditions was far more regulated than in other contexts, but polar life did still provide explorers with opportunities to use their bodies in new and different ways.

CONCLUSION

Studying sex has proved one of those challenging aspects of writing this book. In large part this is a result of the scant information we have available from explorers in published and private sources. I have often found myself constrained by the fact that many explorers dared not write about certain sexual matters, even in unpublished diaries. Such omissions are understandable, as their "private" diaries were often shared with

FIG. 35. "Minstrel Troupe" on the British National Antarctic Expedition. Reproduced with permission of the Scott Polar Research Institute, University of Cambridge. Image ref: P83/6/1/1/80.

others, as they sought to prove the veracity of their claims and publish a travel narrative. Perhaps what is most interesting about studying sex on expeditions is that it brings constant attention to what we do not know and cannot know—reminding us how partial our pictures of explorers' bodies can be. Even in the references to sex that I have found, it is all too easy to lose sight of the real, messy, embodied encounters that such incidents involved.

The limited information we do have, though, still allows us to understand the relationship between sex, travel, and exploration in new ways. We see that many African explorers engaged in intimate sexual and sometimes emotional relationships. These took various forms. Several explorers traveled as couples, in ways that destabilize our understanding of the singular and solitary explorer, and remind us of the involvement of European women in several journeys. We see that expeditions could be the site of lasting relationships as well as emotional and physical intimacy. But explorers who set out without such ties also formed different kinds of relationships on their journeys. These encounters were shaped by power relations and explorers' moral ideas, and the contexts into which they traveled. African explorers' sexual interactions thus ranged from flirtation and romance to disturbing incidents of abuse and exploitation.

While explorers engaged in sexual practices that were taboo at home, they also tried to control and regulate the bodies of African women. In this sense we see how exploration was powerfully influenced by patriarchal ideas: many men sought to portray themselves as unattached while at the same time viewing unmarried women as a moral threat. But we should remember that expedition leaders did not have unlimited power, as women could and did run away from expeditions. The pressures facing white women explorers were different, for they had far more power in colonial contexts. They still had to carefully navigate patriarchal norms and the legal confines of marriage. Avoiding sex (or at least any mention or rumor of it) was vital for protecting their domestic reputation. But they did sometimes write about romance in their accounts.

The sexual bodies of Antarctic explorers were highly regulated in both formal and informal ways. These ranged from pre-expedition medical examinations to the lack of privacy of a polar hut. Yet, even in these contexts, travel still provided explorers with opportunities to use their bodies in new and different ways. Taken together these various relationships show how explorers' sexual bodies were shaped by questions of power and gender while illustrating the limits of understanding them as the emotionally detached "rolling stones" described by Speke at the start of the chapter. Within these encounters we often find travelers failing to maintain any sort of emotional or physical distance from the people around them.

CONCLUSION

Early in this book, I described the bronze statues of David Livingstone and Ernest Shackleton outside the headquarters of the Royal Geographical Society (RGS-IBG) at Kensington Gore, London. I end it by returning to the bodies of these explorers, but from a different perspective, focusing not on their heroic representations but on their digestive systems. Both men experienced problems emanating from their bowels. On his final expedition, Livingstone suffered from acute dysentery and was "bleeding profusely" from his anus in the run-up to his death.[1] After he passed away on May 1, 1873, a man named Farajalllah removed his guts and internal organs, so they could be buried and the body preserved. On opening Livingstone's abdomen, he found "a clot of coagulated blood, as large as a man's hand," which was probably the cause of his death.[2]

Shackleton also seems to have found himself unable to control his guts while traveling. In February 1909 Shackleton, Eric Marshall, Jameson Adams, and Frank Wild were rushing north, having come within ninety-seven miles of the South Pole. They were desperate to reach their ship, the *Nimrod*, before it returned to New Zealand at the start of March. All were extremely hungry and began suffering from dysentery after eating tainted pony flesh, with Wild particularly badly affected. As they approached the coast, Shackleton and Wild went ahead to get help. They eventually reached the hut erected by the British National Antarctic Expedition in McMurdo Sound, near their own base at Cape Royds and the ship. From here they were able to signal and receive assistance from the other men there. Before long all four men were on their way home. This story is evocatively described in Shackleton's book, *The Heart of the Antarctic* (1909), but the published narrative glossed over certain aspects of the men's experiences. When Robert F. Scott arrived at the same location less than a year later, on his own South Pole attempt, he was horrified by the "offensive condition" in which Shackleton had left the hut: "Boxes full of excrement were found near the provisions and filth of a similar description was thick under the veranda & even in the corners of the hut itself. It's extraordinary to think that people could have lived in such a horrible manner."[3]

That Livingstone and Shackleton suffered with such gut troubles is not surprising, but what we should remember is how different the static statues of these two men are from the leaky, messy bodies they had in the field. Nor is the partial image we are left with accidental. Before publication Horace Waller, a fellow missionary and antislavery activist, edited Livingstone's diaries, removing discussion of his hemorrhoids and other frank anatomical descriptions.[4] Scott's comments about the piles of excrement left by Shackleton's expedition were edited out by the committee that posthumously published his diaries. As elsewhere open discussions of "abject" bodily functions were unsuitable for inclusion in a book of exploration.[5] Aspects of bodily experience that might be central to the success of an expedition—such as the state of a travelers' digestive system—could only be discussed in sanitized terms in published accounts. Explorers were also under pressure to avoid writing down certain experiences or incidents in even their private letters and diaries. Already we begin to see the differences between explorers' real, material bodies and public representations of them.

I do not mention these incidents to drag either explorer's name through the muck for the sake of it. Rather I use them to illustrate one of the central arguments of this book: that focusing on the body can allow us to better understand exploration. Travelers' bodies were far from heroic, solid metal statues. They were material, fleshy, and leaky, and dependent on the support and care of others.[6] By focusing on these incidents, we can see exploration in new and different ways, highlighting the multiple bodies involved in travelers' journeys.

Livingstone and Shackleton can additionally help us to understand the enduring appeal of Victorian and Edwardian explorers. The large numbers of explorers, their varied personalities, and their journeys to different regions have allowed commentators and audiences to find alternative heroes within the history of exploration. Despite remaining an admired figure well into the twentieth century (even among some African nationalist leaders), Livingstone is "no longer the household name he once was."[7] In contrast, public interest in Shackleton has surged in the twenty-first century, including through a 2002 miniseries starring Kenneth Branagh and a 2022 expedition that successfully located the wreck of his ship, the *Endurance*, at the bottom of the Weddell Sea.[8] Contemporary retellings of Shackleton's journeys focus less on his 1907–9 attempt to reach the South Pole than on his leadership decisions following the sinking of the *Endurance* on the Imperial Trans-Antarctic Expedition (1914–17). Shackleton is a more palatable hero for twenty-first century audiences because he traveled to unpopulated Antarctica; therefore, he is seen as having less colonial baggage than many other explorers. Indeed, he has made the transition from imperial hero to management guru and remains popular among audiences generally critical of empire and exploration.[9]

Shackleton is not the only explorer who has made a comeback in the twenty-first century. Dan Simmons's 2007 book *The Terror* and the 2018 AMC series based on it have revived interest in the fate of the Franklin Expedition.[10] These works have had a powerful impact because they coincided with a successful search, led by Parks Canada, for the wrecks of the expedition's ships in Nunavut waters. HMS *Erebus* was found in 2014 and HMS *Terror* in 2016. The extraction of these shipwrecks resulted in the recovery of various objects, some of which were exhibited at a new "Death in the Ice" exhibition, initially located at the Canadian Museum of History. The explorers' bodies—and their breakdown through scurvy and starvation as well as their dismemberment through cannibalism—remain a central theme of this exhibition and in fictional twenty-first-century accounts of the expeditions.[11] But these recent interpretations of the Franklin Expedition have, in general, been more attentive to Inuit contributions and histories than the Victorian debates that I began this book by discussing.[12]

In this sense we see that exploration histories remain popular because they can be continually retold in ways that appeal to different audiences and contemporary concerns, as metabiographies of explorers have pointed out.[13] This book might be understood as part of this tradition of continual reinterpretation, focusing on aspects of exploration most important for contemporary researchers seeking to rethink the history of geography.

BODIES, EMBODIED LABOR, AND SUBALTERN HISTORIES

If exploration is likely to remain a part of public culture, then it is worth thinking about how to communicate a more diverse history. This book provides a new framework for conceptualizing these histories through focusing on expeditionary bodies, an approach that allows us to more fully account for the contributions of subaltern people to the history of geography. If we start from analyzing the day-to-day work that made an expedition possible, we begin to see a much more diverse account than that found in traditional exploration histories. One important strand of this work has been to think about the embodied labor that made expeditions possible. This has included the work of cooking, cleaning, and carrying on an expedition as well as of observing, navigating, translating, and instilling discipline within an expeditionary party.

Studying expeditionary bodies also provides a means to think about the work of representation. As I have argued in both chapter 1 and chapter 7, explorers often sought to present their bodies in the most favorable light to domestic audiences. Sometimes this involved covering up intimate sexual encounters or recasting incidents of bodily breakdown. In this way this book prompts us to reflect on the relationship between travel, movement, and writing. Explorers used what Edward Said terms "the power to narrate, or to block other narratives from forming and emerging" to reframe their dependence or vulnerability.[14] But such blocking was never total, in part because exploration brought together various practices that did not always sit alongside each other easily. Although explorers could try to write about their experiences in heroic terms, even singular bodies were understood in multiple ways.

There were also limits on the degree to which explorers could rhetorically manage either bodily failure, illicit sexual intimacy, or anything else. For a start illness could shape the quality and quantity of observations that explorers could make. Falsifying records might be possible (and did, on occasion, happen). But direct falsification was a risky strategy, given the level of scrutiny to which many travelers' diaries were exposed. These constraints were particularly acute in the context of large expeditions in which multiple Europeans kept diaries. In such situations conflicts were common, and explorers often sought to record (and sometimes even publish) information that cast their companions in a negative light.

The importance of individual heroism within exploration cultures contributed to feuds between explorers and to the recording of more problematic aspects of bodily experience. Explorers often sought to suggest that they were stronger and braver than those they traveled with. This was easily done when an explorer was the only one keeping a record of the journey, but problems could emerge when two explorers sought to cast the same event in different ways. Such incidents could lead to broader debates over credibility, which often centered on explorers' bodies and the ways they had behaved and been used.

This book calls into question many of the unhelpful binaries prevalent in explorers' writings. Distinctions between feminine "travel" and more masculine "exploration" seem increasingly hard to maintain. When studied alongside each other, it is hard to see how the activities of Mary Kingsley or Isabella Bird were any less journeys of exploration than those of their male contemporaries. But it is also difficult to maintain rigid distinctions between expedition leaders and their so-called guides or intermediaries. There was seldom a neat separation between the kinds of work done by explorers in one context and designated as the menial or logistical work of servants, valets or guides in another. Terms like "indigenous," "native," or "local" can be equally misleading, as subaltern expedition members were often traveling far away from their own homes. The divergent standards applied in different contexts are visible through the comparison between Antarctic expeditions and those on other continents. Explorers described cooking as an essential—and perhaps even heroic—labor when it was carried out by white (particularly middle- and upper-class) men sledding in Antarctica. Though such work was not always equally valued or celebrated when carried out by working-class sailors, it was still portrayed as an essential part of expeditionary activity. When similar tasks were conducted by African or Asian men and women, those who did it were portrayed as "servants" rather than equal expedition members.

Men such as James Chuma, Ito Tsurukichi, and Sidi Mubarak Bombay performed important managerial work on expeditions and became established (and even famous) travelers—but they seldom earned the title "explorer." The way these different experiences were valued reflected the identity of the people in question and their ability to write about their experiences in books and articles rather than their contribution to the work involved in each journey. We must devote greater attention to how questions of identity, location, and authorship shaped the ways different bodies and activities were appreciated. The idea of the explorer is clearly racially charged and highly gendered, meaning that it can obscure the complex relationship and multiple forms of work that travel involved.

Nor was there a clear division between explorers and the local population: expeditions often relied on, lived with, and formed intimate and exploitative relationships with people they encountered.

Explorers were also not the only ones producing knowledge. As I argued in chapter 6, travelers might also find their own bodies scrutinized and observed, sometimes in invasive and uncomfortable ways. In some situations this curiosity could be important in facilitating explorers' journeys. Meanwhile, focusing on areas where explorers met resistance to the presence of their bodies allows the examination of both anticolonial resistance and the relationship between exploration and more violent forms of colonial expansion.

Taking the body as the central focus also allows us to better include animals in the history of geography. This book therefore offers new insights on the importance of animals in Victorian and Edwardian geographical fieldwork. Collecting, preserving, and transporting animal bodies was an important part of expeditionary labor and one that often depended on the skills and knowledge of subaltern people. Addressing animal bodies brings further attention to the prejudiced nature of explorers' writings and to the hybridity of expeditionary parties.

Expeditions were often "hybrid" undertakings, involving the cooperation and interdependence of multiple species.[15] Animals played various roles, often providing transport and companionship, and served as a source of food. Specially adapted animal bodies were vital in enabling explorers to travel in extreme environments where the human body struggled, including huskies and Siberian ponies in cold or high-altitude environments or camels in the Australian desert. In such situations interspecies relations of care and dependence were central to the success of an expedition. Some explorers were more sensitive to the ethical implications of using animals. Yet there were clearly limits to such relations of care, and expeditionary animals frequently suffered greatly and had high mortality rates. Much writing about expeditionary animals only discusses their usefulness (and relationship) to humans—a limit that it is difficult to overcome, given the evidence available. But we occasionally get glimpses of animal agency and resistance, most clearly illustrated in Scott's writings about his dependence on the disciplinary structures of the huskie pack or his encounters with orcas.

We also need to be attentive to the ways that explorers often anthropomorphized certain animals while using animalizing language to dehumanize and racialize subaltern people. Some animals are more visible in explorers' writings than many of the people they relied on. The history of expeditionary animals cannot be easily separated from studies of nineteenth century racism and empire. Historians working on the more-than-human in this context need to be cautious that their work does not further entrench the erasure of racialized people from the history of the discipline.[16] Animals cannot simply be added to the history of exploration but must be part of a broader conversation about undervalued contributions and experiences.

BODIES AND ARCHIVES

This book considers the nature of exploration records and the challenges of reading them "against the grain."[17] Archival catalogs, for various understandable reasons, tend to privilege names, dates, and locations in their descriptions of documents; consequently, they cannot easily be interrogated to reveal the presence of material specifically addressing the body or subaltern labor. An explorer may write a letter or diary entry covering numerous other topics, before moving on to mention, in passing, the illness of a companion, their own blindness, or what someone else prepared them for lunch. It is unlikely that the document in question will be described in the catalog so as to draw attention to the latter points. Such information makes it far easier to focus on individuals than themes, contributing to the fact that biography remains the dominant mode of writing about explorers. Researching the bodies of all those on expeditions requires sifting through and drawing on a wide range of sources and calling on the knowledge and experience of archivists. Using resources in this way forces us to think about how the acts of cataloging shaped the questions it is possible (or at least straightforward) to research.

Despite such issues a great deal of information can be found about the contributions and experiences of subaltern people in the public and private records of exploration. Explorers were willing to note the presence and roles of such individuals within their expeditions, even in their published accounts. The problem for contemporary readers is that they often did so in such offensive and dismissive terms that distort subaltern people's motivations and minimize their contributions. Explorers' descriptions are permeated by racist caricatures and stereotypes or fail to give information about the identities of people who did important expeditionary work. On Antarctic expeditions class prejudice played a related role in shaping how expedition leaders wrote about those they commanded. Nor are visual sources such as sketches, photographs, and other artwork free from racism. The struggle has not, primarily, been to find relevant sources but to find ways to interpret them without repeating explorers' prejudices. But explorers' writings are not monolithic, and, at certain times and in certain ways, they sometimes give us more information about the contributions and experiences of those they depended on and interacted with. Overall this book brings attention to the different ways and different degrees to which certain voices are silenced by or absent from the public, private, and visual records of exploration.

Explorers wrote differently in public and private. At times "raw" private records provide more honest accounts, as both explorers and their publishers sought to recast travelers' journeys in terms more favorable to domestic audiences.[18] But private writings do not always exist, and, where they do, they are not necessarily any less limited than published sources. In writing this book, I became ever more aware of the limits of viewing the rewriting of accounts only as a negative process of censorship and distortion. Often, as we have seen, various reputationally damaging details were removed from published narratives. But other information might be added. In the field explorers were often under considerable physical and mental strain, meaning that they typically only wrote down particularly interesting or novel events, or those required by dint of routine scientific observation or survey work. Day-to-day domestic arrangements and experiences sometimes receive surprisingly little attention in the raw records of exploration. When undergoing extreme physical suffering, explorers often had little time or energy to write about the experience in detail.[19] Scott explained: "The diaries which record the doings of a sledge party, and which are written in such adverse circumstances . . . do not enter into the hardships and discomforts which are inevitable to the day's work, but in the main are devoted to special incidents of the particular day."[20] As this passage shows, suffering could draw attention away from commonplace activities. Scott's drafts of *The Voyage of the* Discovery (1905) shows him adding supplementary details about the party's experiences of hunger and Petty Officer Edgar Evan's nose and its liability to frostbite.[21] Mary Kingsley reported similar problems when trying to turn her diaries from her West African travels into a book, finding that many important issues were left out.[22] In highlighting the contribution of non-Europeans to the domestic and caring labor of the expedition, chapter 2 relied much more heavily on published sources than I expected. Published accounts could, and often did, serve to write the body back into expedition narratives.

Travel narratives often devoted specific sections of their books to "logistical" questions, summarizing the composition of their expeditions in ways that their diaries did not. Nor are unpublished records free from racism and other prejudices. Private records, in short, are limited in their own ways and present their challenges when it comes to studying explorers' bodies. It is in the act of comparison between the raw and the polished that the most significant insights can be gained. The comparison of written and visual sources can also offer insights into subaltern contributions, as photographs and drawings made by explorers sometimes show subaltern labor and experiences in ways that are missing from their written accounts. Analyzing multiple kinds of sources is key to recovering the work of subaltern people.

Another important approach has been to compare the different ways the same labor was valued in various contexts and when carried out by different people. What counted as heroic or important could vary considerably: the work of cooking, cleaning, moving, caring, and carrying could be treated very differently depending on who was doing it; so, too, could curiosity about the bodies of others or, indeed, sexual desire. In this sense another strand of my methodology has been to place travelers' writings in their various social, political, and cultural contexts.[23] This task has required an understanding of historical developments in the regions through which explorers traveled, drawing on insights from African and Asian history and critical scholarship on racism and empire. Such work has enabled me to piece together a more complete picture of the impact of the ways explorers' writings changed over time and space.

Even after deploying various strategies, the limitations of the sources available sometimes leave us with more questions than answers. We often do not know the exact motivations of people whose voices are only recorded in such fragmentary ways. In some contexts one way to address this disparity would be to conduct oral history interviews in the areas where explorers traveled. In the case of the search for HMS *Erebus* and HMS *Terror*, such oral histories have already proved invaluable in developing our understanding of the Franklin Expedition. Similar methods have the potential to enhance (and perhaps upend) our understandings of explorers and their activities in other contexts. Until we do so, it is hard to replace the old "heroic" stories of exploration with something equally definitive. But there are clearly wild inconsistencies within the standard colonial narrative. We are simultaneously told that explorers went to "unexplored" and "blank" places, yet their accounts are full of encounters with local people with a clear sense of a country's geography. We read that many of their journeys were unprecedented, but their accounts show that they often followed preexisting trade routes. Explorers often tell us that those under their command were childlike, animalistic, or successful only to the degree that they were obedient. But their writings also show that much of the day-to-day discipline of many expeditions was carried out by headmen and intermediaries, and that many of the people they disparaged had high standards of self-discipline and self-control. Explorers' accounts are, in this sense, wildly contradictory, but can be read in ways that surface subaltern contributions.

I have shown that a range of people were involved in British-led geographical expeditions. But what do we gain by focusing on this diversity? The purpose in foregrounding such narratives is not simply to diversify the history of exploration but to examine the challenges of fully accounting for these experiences and contributions. In doing so we see that the history of exploration is colonial in quite fundamental ways, both because of the one-sided evidence available, and because the benefits of exploration overwhelmingly went to metropolitan individuals, governments, and institutions. I have attempted to tell these stories from a different angle, but we should not lose sight of the broader role that exploration played in the expansion of imperialism or the development of colonial systems of knowledge production that still shape how we (both in Britain and formerly colonized countries) study and think about the world.[24] I therefore want to spend some time in the final part of this book reflecting on how understanding exploration differently can advance more contemporary discussions.

DIVERSIFYING HISTORY, RETHINKING GEOGRAPHY

Geography is still grappling with the legacies of exploration and empire. Expeditions embody many of the practices that most contemporary geographers are keen to move away from. In this book I have shown that critically engaging with exploration histories can be an important way of understanding geography's problematic colonial history. These histories continue to shape geography's "spatial present." Images and statues of prominent travelers continue to cover the walls of geographical and academic institutions (such as the headquarters of the RGS-IBG). Such artwork shapes who feels comfortable and uncomfortable in institutional spaces, contributing to the whiteness of the discipline.[25] Part of the process of decolonizing geography should be to visibly and more permanently record the roles that subaltern people played within Victorian and Edwardian expeditions. In some instances using the biographies and artwork of prominent individuals might be strategically useful in starting new conversations, but we should avoid glossing over the experiences and contributions of individuals unnamed in explorers' accounts. Focusing on bodies provides a means to acknowledge their presence and labor within the history of geography.

Rethinking exploration also has implications for how we understand the continuing importance of fieldwork within geography. The discipline has relied on embodied observations made "in the field" more than many other subjects. The importance of such practices has made it a science that has long grappled with the reliability and credibility of bodies and senses. From early on in geography's history, some have privileged embodied observations of mobile bodies over more sedentary forms of research.[26] These imperatives have not disappeared. Fieldwork remains an important part of both geographical teaching and research and is viewed as a necessary "rite of passage" in the geosciences.[27] Thus it is important to reflect on the relationship between bodies and mobility in geography's past and present.

When done "ethically, sustainably and respectfully," fieldwork is a vital part of geographical education and research.[28] But, for some time, feminist geographers have noted how field sites are often constructed as a "masculine domain," in which researchers perform heroic visions of manliness similar to those discussed in this book.[29] Women still face additional challenges in the field, including gender-based violence and sexual harassment. But the idea of the field as a masculine space has long been contested. Even in the Victorian era, white women found ways to engage in geographical research—though, as we have seen, they often faced additional challenges when they did so.[30]

More recent critiques of fieldwork have devoted sustained attention to other people often disadvantaged by current fieldwork practices, including people of color, LGBTQ, and working-class geographers as well as researchers with mental health issues or disabilities.[31] Issues such as "expense, physical exertion, alcohol culture, and access to toilets" make fieldwork inaccessible for minoritized groups. Many of these problems have their origins in the masculine, heroic vision of exploration critiqued in this book. Scholars and institutions have begun to address some of these challenges, but there is still more to do in ensuring that field sites are accessible to all.[32] It is also worth reflecting on broader inequalities in who conducts fieldwork and where it is conducted. Due to the geopolitical legacies of empire, it remains far more difficult for scholars from the Global South to conduct research in the Global North than the other way around. These broader problems cannot be fully addressed by individual institutions and require government action.

There are important similarities as well between the modes of exploration critiqued in this book and forms of "parachute science." This practice involves scholars from the Global North traveling to countries in the Global South to conduct research without collaborating with or "recognising the participation of local researchers, landowners, or guides."[33] Parachute research can also involve the extraction of items and samples without the collaboration or consent of local populations. Such types of fieldwork continue to rely on forms of "body work" that are rarely fully acknowledged, including that of guides, cooks, drivers, local populations, and translators. Contemporary research by northern scholars in the Global South can replicate many of the practices of dependence and erasure critiqued in this book. Moving away from parachute science requires a greater acknowledgment of the embodied, sensory, and caring labor that makes field research possible.

But there is also a danger of locating all geography's problems in "the field." As Lioba Hirsch has demonstrated, such an approach is often based on the assumption that the university is "a stable, rational place of knowledge production and management" and a place of safety for students and staff. Meanwhile, "the field," particularly when located in formerly colonized countries, is understood as "a place of knowledge extraction, often perceived as unstable, violent and irrational."[34] This binary, Hirsch demonstrates, obscures the discrimination and harassment that women, religious minorities, and people of color experience within contemporary academic institutions in the Global North.

We must also avoid presenting geography's troubles as the product of a distant colonial past. Without critical reflection on the present, studying the history of exploration could function as a "move of innocence," assuaging white guilt and drawing attention away from the coloniality of contemporary institutions.[35] Universities, for many, remain oppressive structures, imbued with the legacies of racism and empire.[36] Academic British geography, for instance, remains an overwhelmingly white discipline, extremely so at the professorial level.[37]

But focusing on the body raises broader questions about how we value embodied labor. The history of exploration shows us that our minds and bodies cannot be separated. Explorers' bodies were continually implicated in the knowledge they produced. Illness, starvation, and exhaustion often left lasting marks on the quality of the findings they were able to bring back. When they were able to produce reliable knowledge, it is because they had so many of their bodily needs met by others.

Today, research still depends on the embodied labor of numerous workers—cleaners, cooks, and administrators—who are seldom referred to as geographers or academics and who are frequently paid much less than those engaged in more cerebral pursuits.[38] Much care work is also unpaid and carried out at home. The production of knowledge remains a collaborative process, relying on multiple bodies and various forms of labor. Despite this the myth of the lone researcher is perpetuated by academic performance measures, tenure requirements, and promotion criteria that value monographs and single-author publications. In this sense, many of the tools I have developed could also be applied to contemporary institutions, and, indeed, to the production of this book.

NOTES

INTRODUCTION

1. Woodman, *Unravelling the Franklin Mystery*. The exact timing is a source of debate. Woodman argues that some survived beyond the 1848 abandonment of the ships.
2. May and Beattie, "Evidence for End-Stage Cannibalism."
3. Cavell, *Tracing the Connected Narrative*, 202–29.
4. Charles Dickens, "The Lost Arctic Voyagers," *Household Words*, December 2, 1854, 365.
5. John Rae, "Dr. Rae's Report," *Household Words*, December 30, 1854, 249.
6. Cavell, "Publishing Sir John Franklin's Fate," 160.
7. For examples, see J. I. MacNair, "Some Recovered Relics of David Livingstone," JMS/21/148, RGS Archives; "Framed Notice Relating to the Franklin Expedition Relics from the 1891 Royal Naval Exhibition at Chelsea," National Maritime Museum, AAB0506.
8. Driver, "Exploration as Knowledge Transfer," 94.
9. Driver, "Exploration as Knowledge Transfer"; Driver "Hidden Histories Made Visible?"; Driver, "Face to Face with Nain Singh"; Kennedy, *Last Blank*; Jones, "Local Knowledge and Indigenous Agency"; Wisnicki, *Fieldwork of Empire*; Fleetwood, *Science on the Roof of the World*; Shellam et al., *Brokers and Boundaries*.
10. See, for example, Sivasundaram, "Prehistory of COVID-19"; Sivasundaram, "Imperial Transgressions"; Saha, *Colonizing Animals*; and Swart, *Riding High*.
11. Livingstone, *Geographical Tradition*.
12. For an introduction to these debates, see "Decolonising Geographical Knowledges," themed intervention, *Transactions of the Institute of British Geographers* 42, no. 3; and Radcliffe, *Decolonizing Geography*.
13. Craggs, "Decolonising *The Geographical Tradition*'" see also Schelhaas et al., *Decolonising and Internationalising Geography*.
14. Griffiths and Baker, "Decolonising the Spaces," 455–58; Kearns, "Topple the Racists 2"; Jazeel et al., "Decolonising Spaces."

15. Kearns, "Topple the Racists 1."
16. Puttick and Murrey, "Confronting the Deafening Silence"; Johnson, "Throwing Our Bodies," 89–96.
17. Bracken and Mawdsley, "Muddy Glee," 280–86; Leyland et al., "Classics Revisited."
18. Guasco, "On the Ethics of Not Going There"; Howkins, "Have You Been There?"
19. Kennedy, "Introduction," in *Reinterpreting Exploration*, 3–4. See also Bridges, "Explorers' Texts," 67; and Kennedy, *Last Blank*, 64. For examples of popular biographies, see Jeal, *Livingstone*; Jeal, *Stanley*; and Fiennes, *Captain Scott*.
20. Coupland, *Livingstone's Last Journey*, 7.
21. Baigent, "Deeds Not Words?"
22. Brodie, *Devil Drives*; Huntford, *Last Place on Earth*; Jeal, *Livingstone*.
23. For example, Livingstone, *Livingstone's "Lives"*; Rupke, *Alexander Von Humboldt*; Withers, *Majestic River*. On this approach in general, see Saunders, "Defining Metabiography."
24. Said, *Orientalism*; Said, *Culture and Imperialism*; Carter, *Road to Botany Bay*; Pratt, *Imperial Eyes*. See also Gregory, *Geographical Imaginations*.
25. "Imaginative geographies" comes from Said, *Orientalism*, 49; the second quote is from Pratt, *Imperial Eyes*, 3.
26. Regard, introduction to *British Narratives of Exploration*, 10; emphasis in original.
27. Kennedy, *Imperial History Wars*, 61.
28. Driver, *Geography Militant*, 8.
29. Withers, "Trust—in Geography."
30. Shapin, *Social History of Truth*, 245.
31. Kennedy, *Last Blank*, 195–232; Fabian, *Out of Our Minds*.
32. Fleetwood, "No Former Travellers."
33. Keighren et al., *Travels into Print*, 99, 75.
34. Keighren et al., *Travels into Print*, 82–92, 96; quote at 92.
35. Riffenburgh, *Myth of the Explorer*.
36. Livingstone, *Putting Science in Its Place*, 153; Driver, *Geography Militant*, 49–67; Fabian, *Out of Our Minds*, 78.
37. Thompson, *Suffering Traveller*.
38. Livingstone, *Putting Science*, 153; Kennedy, *Last Blank*, 94.
39. Outram, "On Being Perseus," 290.
40. MacDonald and Withers, "Introduction," 2.
41. Warrior and McAleer, "Objects of Exploration," 117.
42. Ryan, *Photography and Exploration*; Withers, "Geography and 'Thing Knowledge.'"

43. Fleetwood, "No Former Travellers," 7. See also Driver "Distance and Disturbance," 73–92; on instrument failure and repair, see Schaffer, "Easily Cracked."
44. Michael, "Making Mount Kilimanjaro German"; Withers, "Geography and 'Thing Knowledge.'"
45. Driver, "Face to Face with Nain Singh"; Driver et al., *Mobile Museums.*
46. Dritsas, "From Lake Nyassa to Philadelphia," 50.
47. McCook, "It May Be Truth."
48. Heffernan, "Dream as Frail." See also Hevly, "Heroic Science."
49. Driver and Jones, *Hidden Histories*, 5.
50. Driver and Jones, *Hidden Histories.* The exhibition has since been published online. Royal Geographical Society (with IBG), "Hidden Histories of Exploration," accessed April 3, 2024, https://www.rgs.org/our-collections/stories-from-our-collections/online-exhibitions/hidden-histories-of-exploration.
51. Jones, "Local Knowledge and Indigenous Agency," 14; Driver and Jones, *Hidden Histories.*
52. For example, Kennedy, *Last Blank*; Fleetwood, *Science on the Roof*; Konishi et al., *Indigenous Intermediaries*; Jones, "Local Knowledge and Indigenous Agency"; and Schaffer et al., *Brokered World.*
53. Fleetwood, *Science on the Roof*; Kaalund, *Explorations in the Icy North*; Lemtur, *Locating Himalayan Porters*; Martin, "Indigenous Tales of the Beaufort Sea"; Martin, "'Deer-Men' and the 'Bowhead-Men'"; Wisnicki, *Fieldwork of Empire.* See also Brescius, *German Science in the Age of Empire*; Simpson, *Frontier in British India*; and Heggie, *Higher and Colder.*
54. For example, Haraway, *Companion Species Manifesto*; and Haraway, *When Species Meet.*
55. Saha, *Colonizing Animals*; Saha, "Colonizing Elephants"; Swart, *Riding High.* See also Burton and Mawani, *Animalia*; Cribbet et al., *Wild Man from Borneo*; and Skabelund, "Animals and Imperialism."
56. On the more-than-human in geography, see Morin, "Carceral Spaces"; Urbanik, *Placing Animals*; Oliver, "Beyond-Human Research"; and Philo and Wilbert, *Animal Spaces.* On expeditionary animals, see Forsyth, "More-than-Human Geographies"; Oliver, "Animals of the Royal Geographical Society"; and Wilks, "Coldest Dog and Pony Show."
57. Cox, "Armchair Geography"; Lambert, *Mastering the Niger.*
58. Mayhew, "Halford Mackinder's 'New' Political Geography"; Stoddart, "Geography, Education, Research."
59. Driver, "Active Life."
60. George Curzon, "Ladies and The Royal Geographical Society," *Times*, May 31, 1893, 11.

61. Lord Curzon, president of the Council, letter to Council on Subject of Women's Admission to the RGS, October 24, 1912, AP93/6, RGS Archives, fols. 2–3.
62. Leshem and Pinkerton, "Rethinking Expeditions"; Williams and Love, "Low-Carbon Research."
63. The following two paragraphs have been published previously in modified form in Armston-Sheret, "Diversifying the Historical Geographies." Reprinted by permission of the author and publisher.
64. Noxolo, interview with GeogPod, accessed April 3, 2024, https://geography.org.uk/geogpod-the-gas-podcast.
65. Pratt, *Imperial Eyes*, 9.
66. Said, *Culture and Imperialism*, xiii.
67. Kennedy, *Last Blank*, 1–24.
68. Robinson, "Scientific Travel."
69. Craciun, "What Is an Explorer?," 30.
70. Craciun, "What Is an Explorer?," 31; Fernandez-Armesto, *Pathfinders*.
71. Kennedy, "Introduction," 2.
72. Craciun, "What Is an Explorer?," 30.
73. Davis, *Into the Silence*; Jones, *Last Great Quest*, 269–83. There were, of course, some continuities, too; see Naylor and Ryan, *New Spaces of Exploration*.
74. Jones, *Last Great Quest*.
75. Robinson, "Shackleton Syndrome."
76. Driver, *Geography Militant*.
77. Driver, *Geography Militant*, 117–45.
78. Shapin, "Placing the View from Nowhere," 8. On this issue, see Driver, "Distance and Disturbance"; McCook, "It May Be Truth"; and Withers, "Trust—in Geography."
79. Shapin, *Social History of Truth*, 245.
80. Kennedy, *Last Blank Spaces*, 76; for a similar argument, see Fleetwood, *Science on the Roof*.
81. Green, "Rethinking the Subaltern."
82. Green, "Rethinking the Subaltern."
83. Guha, *Subaltern Studies 1*; Guha, *Subaltern Studies Reader*.
84. Spivak, "Can the Subaltern Speak?"
85. O'Hanlon, "Recovering the Subject," 222; Cowan, *Subaltern Frontiers*, 19. The use of "subaltern" in this way is consistent with earlier scholarship. Guha conceptualizes "subalternity" as a status produced through the relationship between elite and dominated groups. Guha, *Elementary Aspects*, 64–65, 75.
86. Jazeel, "Subaltern Geographies," 88–103.
87. Chakrabarty, "Small History of Subaltern Studies," 472.

88. Birkett, *Spinsters Abroad*; Middleton, "Some Victorian Lady Travellers."
89. Maddrell, *Complex Locations*, 5. After her marriage in 1881, she often went by Mrs. J. Bishop or Isabella Bird Bishop, and she published under these names. For reasons of clarity and consistency, I have chosen to use her maiden name throughout the book.
90. Stoddart, "Feminist Historiography of Geography," 484; for a similar argument, see Kennedy, *Last Blank*, 65–66.
91. Stoddart, "Feminist Historiography of Geography," 484.
92. Isabella L. Bird, "Diary-letters, or 'Circular' 24 Dec 1889–15 Jul 1890," MS. 42023, National Library of Scotland (NLS), fols. 25, 27, 28.
93. Johnson, "Illustrating Nature"; Johnson, "On the Colonial Frontier"; Keighren, "Royal Geographical Society for Ladies."
94. Thomas, "What Is an Expedition?," 17–20.
95. Thomas, "What Is an Expedition?"
96. Burton, "Lake Regions," 16.
97. Dritsas, *Zambesi*, 3.
98. Pratt, *Imperial Eyes*; Said, *Orientalism*; Said, *Culture and Imperialism*; Tuhiwai Smith, *Decolonising Methodologies*.
99. Jazeel, *Postcolonialism*, 48.
100. MacKenzie, "David Livingstone." Some travelers also became critical of empire after witnessing anticolonial resistance; see Gopal, *Insurgent Empire*.
101. Hochschild, *King Leopold's Ghost*.
102. Driver, "Henry Morton Stanley and His Critics."
103. Kennedy, *Last Blank*.
104. Spivak, "Can the Subaltern Speak?"
105. Kennedy, *Last Blank*, 251.
106. There are some important examples of African voices being more directly represented in colonial archives. One More Voice, accessed February 28, 2024, https://onemorevoice.org.
107. Driver, "Material Memories of Travel"; Driver and Jones, *Hidden Histories*; Ryan, *Photography and Exploration*.
108. Said, *Culture and Imperialism*, xiii.
109. Driver, "Intermediaries and the Archive"; Wisnicki, *Fieldwork of Empire*, 6.
110. Barrett-Gaines, "Travel Writing," 53; Driver, "Intermediaries and the Archive," 17.
111. Driver, "Intermediaries and the Archive," 17. On reading against the grain, see Stoler, *Along the Archival Grain*, 47.
112. Raj, *Relocating Modern Science*, 64.
113. For an introduction to these literatures, see Shilling, *Body and Social Theory*; and Turner, *Body and Society*.

114. Canning, "Body as Method?," 499–513; Burton, "Body in/as World History," 276–77; Ballantyne and Burton, "Introduction," in *Bodies in Contact*, 4. On using the body to surface subaltern contributions, see Bressey, "Surfacing Black and Brown Bodies."
115. For example, Ballantyne Burton, *Bodies in Contact*; Forth, "Fat, Desire, and Disgust"; Harris, *Punishing the Black Body*; Sinha, *Colonial Masculinity*; String, *Fearing the Black Body*; and Topdar, "Corporeal Empire."
116. Grosz, *Volatile Bodies*, 6.
117. Butler, *Gender Trouble*, 17.
118. Shilling, *Body and Social Theory*, 37–61, 62–87.
119. Robinson, *Lost White Tribe*.
120. Graeber, *Debt*, 244; Hokowhitu et al., *Routledge Handbook of Critical Indigenous Studies*.
121. Butler, *Bodies That Matter*; Butler, *Gender Trouble*; for an alternative perspective, see Grosz, *Volatile Bodies*. These debates have also played out within human geography; see Longhurst, *Bodies*; and Nelson, "Bodies (and Spaces) Do Matter."
122. Quote from Porter, "History of the Body," 215. See also Butler, *Gender Trouble*; Foucault, *Discipline and Punish*; and Shilling, *Body and Social Theory*, 8–9.
123. Bynum, "Why All the Fuss?," 1.
124. Burton, "Body in/as World History," 277. Shilling, *Body and Social Theory*, 9–10, advocates for a similar approach.
125. Mol, *Body Multiple*.
126. Gimlin, "What Is 'Body Work'?"; McDowell, *Working Bodies*.
127. Driver, "Intermediaries and the Archive of Exploration," 12.
128. Pratt, *Imperial Eyes*, 9.

1. "HEROIC" BODIES

1. Ardeleanu, "British Meat Cannery."
2. Beesly, *Sir John Franklin*, 11.
3. Cubbitt and Warren, *Heroic Reputations and Exemplary Lives*; Jones et al., "Decolonising Imperial Heroes"; Jones et al., *Decolonising Imperial Heroes*.
4. Sèbe, *Heroic Imperialists in* Africa; Berenson, *Heroes of Empire*, 28; Driver, *Geography Militant*, 70; Kennedy, *Last Blank Spaces*, 233–60; Livingstone, *Livingstone's "Lives,"* 102–3, 190, 275; McCorristine, *Spectral Arctic*, 35.
5. "100 Great British Heroes."
6. Hill, *White Horizon*; McCorristine, *Spectral Arctic*, 31–39.
7. Chantiluke et al., *Rhodes Must Fall*.

8. Griffiths and Baker, "Decolonising the Spaces"; Jazeel et al., "Decolonising Spaces."
9. Jones, "What Should Historians Do?," 440.
10. Campbell, *Hero with a Thousand Faces.*
11. Jones, "What Should Historians Do?," 441.
12. Riffenburgh, *Myth of the Explorer.*
13. Carlyle, *On Heroes.*
14. Jones, *Last Great Quest.*
15. Jones, "What Should Historians Do?," 446.
16. Jones et al., "Decolonising Imperial Heroes," 778–79. See also, Sèbe, *Heroic Imperialists*; Berenson, *Heroes of Empire*; and Sèbe, "Making of British and French Legends."
17. Jones, "Exploration, Celebrity," 108.
18. Barczewski, *Heroic Failure*, 11.
19. Middleton, "Search for the Nile Sources."
20. Kennedy, *Last Blank Spaces*, 263–65.
21. Kennedy, *Highly Civilized Man*, 58–92.
22. Maitland, *Speke*, 9–11.
23. Kennedy, *Highly Civilized Man*, 88.
24. Lovell, *Rage to Live*, 172–80. For Burton and Speke's conflicting accounts of the incident, see Burton, *First Footsteps*, 95–106; and Speke, *What Led to the Discovery*, 129–48.
25. Lovell, *Rage to Live*, 172–80; Maitland, *Speke*, 34–40.
26. Rockel, "Decentering Exploration in East Africa," 172–94.
27. Burton, *Lake Regions of Central Africa*; Moorehead, *White Nile*, 35.
28. John Hanning Speke to Norton Shaw, July 2, 1858, John Hanning Speke Collection, RGS/JHS/1/2, Royal Geographical Society (with the Institute of British Geographers) Archives (hereafter RGS Archives), fol. 3.
29. Keighren et al., *Travels into Print*, 14–15.
30. Kennedy *Highly Civilized*, 58; Godsall, "Fact and Fiction." Both question how successful his use of disguise was.
31. Lovell, *Rage to Live*, 52, 85; Burton, *Personal Narrative*, 2:415, in footnote.
32. Kennedy, *Highly Civilized*, 67; Lovell, *Rage to Live*, xvii. Later in life Burton claimed to speak twenty-nine languages.
33. Unsigned review of *Personal Narrative of a Pilgrimage to El Medinah and Meccah*, by Richard F. Burton, *Eclectic Review*, August 1855, 217; unsigned review of *Personal Narrative of a Pilgrimage to El Medinah and Meccah* by Richard F. Burton, *New Quarterly Review and Digest of Current Literature, British, American, French, and German* April 1856, 138–46.

34. Kennedy, *Highly Civilized Man*, 65; Lovell, *Rage to Live*, 186. On the issue of disguise, see Keighren et al., *Travels into Print*, 15; and Withers, "Disguise."
35. Kennedy, *Highly Civilized Man*, 65. Even more recent scholarship has suggested that his use of disguise was more than instrumental. See Regard, "Fieldwork as Self-Harrowing"; Kennedy, *Highly Civilized Man*, 58–92; and Said, *Orientalism*, 195–96.
36. Withers, "Disguise," 8n19. Similar problems emerged for other travelers; see McCook, "It May Be Truth," 190.
37. Burton, *Life*, 1:167, 170; Lovell, *Rage to Live*, 49, repeats this claim.
38. Richards, *Sketch of the Career*, 33.
39. Blunt, *My Diaries*, 546; Kennedy, *Highly Civilized Man*, 14.
40. Lovell, *Rage to Live*, 190; Richards, *Sketch of the Career*, 33; Stisted, "Reminiscences of Sir Richard Burton," 336. Each author gives a slightly different explanation of how he received the wound that caused the scar.
41. Wyndham-Quin, *Past Times and Pastimes*, 1:178; this quote was brought to my attention by Brodie, *Devil Drives*, 15.
42. Blunt, *My Diaries*, 544–45, 46.
43. Newman, *Paths without Glory*, 124; Young, *Search for the Sources*, 147.
44. Stisted, *True Life*, 65.
45. Johnston, *Nile Quest*, 130, 129.
46. Burton, *Zanzibar*, 2:374.
47. Speke, *Journal*, ix.
48. Burton, *Zanzibar*, 2:381; Speke, *What Led to the Discovery*, 23.
49. Burton's marginal comments in John Hanning Speke, *What Led to the Discovery of the Source of the Nile* (London: 1864), Richard Francis Burton Library Manuscripts Collection, RFB1680, Huntington Library, 314. Burton objected to this claim, commenting "rot" next to it in his personal copy.
50. Speke, *What Led to the Discovery*, 314.
51. Speke, *Journal*, 288.
52. Speke, *Journal*, 284; see also Speke, *Journal*, 179–80, 186, 284, 288.
53. Jeal, *Explorers of the Nile*, 155; Humphries, *Search for the Nile's Source*, 105.
54. MacQueen, "Captain Speke's Discovery," 106.
55. Rowbotham, "Soldiers of Christ?," 30; Mills, *Discourses of Difference*, 77.
56. Bird, *Lady's Life*, 4th ed., vii.
57. Bird, *Lady's Life*, 1st ed., 9; all quotes are from this edition unless otherwise stated.
58. "New Books," *Times*, November 21, 1879, 3.
59. "A Lady's Life in the Rocky Mountains," *Saturday Review*, November 12, 1879, 604.
60. Stanley "Skirting the Issues."

61. Bird to Murray, November 21, 1879, MS 42024, NLS, fols. 143–44.
62. Bird to Murray, November 21, 1879, MS. 42024, NLS, fol. 145.
63. Bird, *Lady's Life*, 4th ed., vii–iii.
64. Kingsley, *Travels in West Africa*, 622.
65. Blunt, *Travel, Gender, and Imperialism*, 77–78.
66. "Miss Kingsley's Travels in the French Congo," *Standard*, January 25, 1897, 4.
67. "Mary H. Kingsley," *Pall Mall Gazette*, June 6, 1900, 3.
68. Chang, *Britain's Chinese Eye*, 166–67.
69. Bishop, *Yangtze Valley*, 242; Stoddart, *Life of Isabella Bird*, 350.
70. Isabella Bird to John Murray, 1899, before September 2, NLS, MS. 42029, fols. 61–62; on China in British culture, see Fiske, "Orientalism Reconsidered."
71. Bishop, *Yangtze Valley*, 242.
72. Isabella L. Bird to John Murray, July 27, 1877, MS. 42024, NLS, fol. 86.
73. Speke, *What Led to the Discovery*, 224–26.
74. Speke, *What Led to the Discovery*, 148–49.
75. Speke, *What Led to the Discovery*, 149; on diet in Victorian Britain, see Gregory, *Of Victorians and Vegetarians*, 70–97, 190.
76. Henry Morton Stanley to George Grey, January 9, 1892, GD S47.1, George Grey Collections, Auckland Public Library, fol. 6.
77. Thompson, *Suffering Traveller*.
78. Burton, *Lake Regions*, 1:405; for other instances of saintlike suffering, see Burton, see *Lake Regions*, 1:84.
79. Burton, *Life*, 1:330.
80. Thompson, *Suffering Traveller*, 171.
81. Speke, *What Led to the Discovery*, 165, in footnote; John Hanning Speke to Norton Shaw, July 2, 1858, RGS/JHS/1/2, RGS Archives, fol. 1.
82. Speke, *What Led to the Discovery*, 165.
83. Hamilton, *Nile Quest*, 131; Speke, *What Led to the Discovery*, Huntington Library RFB 1680, 165, footnote.
84. Speke, *What Led to the Discovery*, Huntington Library RFB 1680, 187.
85. Burton, *Lake Regions*, 1:222. See also Carnochan, *Sad Story*, 53.
86. Burton, *Lake Regions*, 1:213.
87. Robinson, "Shackleton Syndrome."
88. Bird, *Lady's Life*, 109, 116.
89. Unsigned book review of *A Lady's Life in the Rocky Mountains*, by Isabella Bird, *Athenaeum*, December 27, 1879, 846.
90. Morin, *Frontiers of Femininity*, 60.
91. "Lady's Life," 604; see, for example, Bird, *Lady's Life*, 68–69.
92. Bishop, *Yangtze Valley*, 33.

93. On this issue, see Palmer, "Emily Katharine Bates," 32.
94. Isabella Bird to John Murray, January 27, 1898, NLS, MS. 42028, fols. 95–96.
95. Bird, *Unbeaten Tracks*, 1:vii.
96. Stoddart, *Life of Isabella Bird*, 262.
97. "Obituary: Mrs. Bishop," *Geographical Journal*, November 1904, 596.
98. Davenport, *Celebrated Women*, 437.
99. "Obituary: Mrs. Bishop," *Edinburgh Medical Journal*, November 1904, 383.
100. "Review: *The Life of Isabella Bird* by Anna M. Stoddart," *Geographical Journal*, April 1907, 452–53.
101. "Obituary: Mrs. Bishop," 383.
102. Stoddart, *Life of Isabella Bird*, 260–61.
103. Quote Hill, *White Horizon*, 31; see also McCorristine, *Spectral Arctic*, 31–39.
104. Nares, *Voyage to the Polar Sea*, 1:374.
105. John Lush, 234 Parl. Deb. 3rd series (1877) 1983.
106. For example, "Alcohol and Scurvy," *Lancet*, December 2, 1876, 795; Rae, "Scurvy and the Arctic Expedition," *Lancet*, November 18, 1876, 733; and "Alcohol and Lime Juice in the Arctic Regions," *British Medical Journal*, December 2, 1876, 724.
107. John Rae to Charles Daly, March 9, 1877, American Geographical Society, ACI, box 126, folder 30, fol. 2.
108. *Report of the Admiralty Committee*, xxiii.
109. Lyon Playfair, 234 Parl. Deb. 3d series (1877) 1974.
110. Forth, "Surviving our Paradoxes"; Forth, *Masculinity in the Modern West*, 141–68; Pick, *Faces of Degeneration*; Reidy, "Mountaineering, Masculinity, and the Male Body"; Zweiniger-Bargielowska, *Managing the Body*.
111. Armston-Sheret, "Tainted Bodies." See also Forth, *Masculinity*, 142; and Robinson, "Manliness and Exploration."
112. Zweiniger-Bargielowska, *Managing the Body*, 64–65; Larson, *Empire of Ice*, 147–50. See also Robb, "Way of All Flesh."
113. Fleetwood, "Bodies in High Places," 509.
114. Some of the ideas in the following section have been published previously; see Armston-Sheret, "Tainted Bodies."
115. "The Voyage of the Discovery," March 30, 1903, *Scotsman*, 8.
116. "Scurvy on the *Discovery*," *British Medical Journal*, April 4, 1903, 807.
117. "Scurvy on the *Discovery*."
118. "Scurvy on the *Discovery*."
119. "Farthest South," *Daily Mail*, March 27, 1903, 5.
120. "Farthest South."
121. Clements Markham, "The First Year's Work of the National Antarctic Expedition," *Geographical Journal*, July 1903, 19.

122. On this issue, see Herzig, *Suffering for Science*.
123. "There Can Be Little Doubt . . . ," *Standard*, May 15, 1902, 6.
124. "The Week," *Speaker*, September 17, 1904, 550.
125. "The Week."
126. "Rescued from Antarctic Ice," *Evening Telegraph*, April 2, 1904, 3.
127. Armston-Sheret, "Tainted Bodies."
128. Scott, *Voyage*, 2:329.
129. Scott, *Voyage*, 2:320. See also "Antarctic Explorers," *Evening News*, September 12, 1904, 2.
130. Armston-Sheret, "Tainted Bodies"; Barczewski, *Antarctic Destinies*, 43–45.
131. Edward Evans to Albert Markham, February 13, 1913, MRK/46, Albert Markham Collection, National Maritime Museum Archive, fol. 19.
132. Jones, *Last Great Quest*, 126.
133. Jones, *Last Great Quest*, 102.
134. Scott, *Last Great Quest*, 163.
135. Quoted in George Seaver, "Foreword," in Cherry-Garrard, *Worst Journey in the World* (1994), xxxiv.
136. Bloom, *Gender on Ice*, 127.
137. "Provisions for the Terra Nova," n.d. [1911] in Official Papers Relation to the Antarctic Expedition under Captain Scott, Joseph James Kinsey Collection, Alexander Turnball Library, National Library of New Zealand, fol. 2.
138. Quoted in Seaver, "Foreword."
139. Hill, *White Horizon*, 30, 46.
140. Spufford, *I May Be Some Time*, 5.
141. The image was originally used as a frontispiece in Scott, *Voyage*, but a variant was also used in Scott, *Scott's Last Expedition*.
142. Spufford, *I May Be Some Time*, 37.
143. Scott, *Journals*, 422.
144. Jones, *Last Great Quest*, 10.
145. Ryan, *Photography and Exploration*, 54–55.
146. Lewis-Jones, *Face to Face*.
147. Forth, *Masculinity in the Modern West*, 160.
148. Edward Wilson, "Journal November 1, 1911 to February 27, 1912," MS 1043; BJ, Scott Polar Research Institute [hereafter SPRI], entry January 7, 1912, fol. 204; Longhurst, *Bodies*.
149. Spufford, *I May Be Some Time*, 169.
150. Admiral Beaumont quoted in Jones, *Last Great Quest*, 155.
151. Kennedy, *Last Blank Spaces*, 242.
152. "The Burke and Wills Monument in Melbourne," *Illustrated London News*, July 22, 1865, 76.

153. Cathcart, *Starvation in a Land of Plenty*, 2; Hadwen, "Literature of Burke and Wills."
154. "From Our Australian Correspondent," *Dundee Courier*, January 30, 1862.
155. Pratt et al., "Burke & Wills Commission Report of the Commissioners," 5.
156. Kennedy, *Last Blank Spaces*, 77.
157. MacKenzie, "Iconography of Exemplary Life," 84–104; Jones, *Last Great Quest*, 25. On the role that a heroic Livingstone did play, see Lewis, *Empire of Sentiment*.
158. Jones, *Last Great Quest*, 24; see also McCorristine, *Spectral Arctic*, 35.
159. Jeal, *Livingstone*.
160. Livingstone, "'Body' of Evidence," 20, 18.
161. Livingstone, "'Body' of Evidence," 5.
162. Jones, "From 'Noble Example'"; Jeal, *Livingstone*; Barczewski, *Antarctic Destinies*.
163. Kennedy, *Last Blank Spaces*, 233–60.
164. Baigent, "'Deeds Not Words?,'" 30, 32.
165. Esson et al., "The 2017 RGS-IBG Chair's Theme."
166. Griffiths and Baker, "Decolonising the Spaces."
167. Stoler, *Along the Archival Grain*, 234.
168. Jones, "Local Knowledge," 63.
169. These next two paragraphs are similar to those previously published in Armston-Sheret, "Diversifying the Historical Geographies."
170. Quoted Speke, *What Led to the Discovery*, 212.
171. Speke, *What Led to the Discovery*, 350.
172. Jones, "Local Knowledge and Indigenous Agency"; Driver, "Exploration as Knowledge Transfer."

2. DEPENDENT BODIES

1. Barczewski, *Antarctic Destinies*.
2. For example, Driver, "Intermediaries and the Archive"; Simpson, *Dark Companions*; Driver, *Hidden Histories*; and Jones, "Intermediaries and the Archive." This chapter is a development of the ideas published as Armston-Sheret, "Diversifying the Historical Geographies," reprinted with permission of the author and publisher.
3. Driver, "Hidden Histories," 423.
4. For example, Konishi et al., *Indigenous Intermediaries*. See also Driver and Jones, *Hidden Histories*; Jones, "Local Knowledge"; Kennedy, "Lost in Place"; Martin, "Indigenous Tales"; Martin, "'Deer-Men' and the 'Bowhead-Men'"; and Schaffer et al., *Brokered World*.
5. Driver, "Intermediaries and the Archive," 15–16.

6. Driver, "Face to Face with Nain Singh"; Fleetwood, *Science on the Roof*; Jones, "Local Knowledge"; Kennedy, *Last Blank Spaces*; Lemtur, "Locating Himalayan Porters."
7. Driver, "Intermediaries and the Archive," 25; Driver, "Hidden Histories," 429; Anthony, *Hoosh*, 16; Smith, *Unsung Hero*; Foley, *Crean*; For the problems with this approach, see Maddison, *Class and Colonialism*, 6–7.
8. Driver, "Hidden Histories," 429; Driver, "Exploration as Knowledge Transfer," 94.
9. Twigg, "Carework as a Form of Bodywork," 389. More broadly, see Gimlin, "What Is 'Body Work'?"; and McDowell, *Working Bodies*.
10. Kennedy, *Last Blank Spaces*, chapter 5. Rockel, on whom I will draw, has examined the work of the East African caravan system but largely for the insights it offers on African history, rather than on exploration.
11. Rockel, "Decentering Exploration."
12. Burton, *Zanzibar*, 2:162–63.
13. Rempel, "No Better than a Slave," 281, 286.
14. Laing, *Tippu Tip*, 236–42, discusses the reasons he struggled to hire porters.
15. Stanley, *In Darkest Africa*, 1:476–7.
16. William Bonny, "Autograph Journal of the Rear Column, September 7, 1888–April 8, 1889," HA.01.0098.71, Stanley Archives, Royal Museum for Central Africa, November 10, 1888, 57.
17. Speke to Secretary RGS, May 12, 1861, Large Print Copy, RGS-IBG Archives, JMS 3/53, fol. 2.
18. Burton, "Lake Regions," 16.
19. Grant, *Walk across Africa*, 42–43; see also Speke's reports to the RGS, JMS 3/53, RGS Archives.
20. Rockel, *Carriers of Culture*.
21. Burton, "Lake Regions," 16.
22. For example, Burton, *Lake Regions*, 2:146, 161–62.
23. Rempel, "No Better than a Slave."
24. Stanley, *How I Found Livingstone*, 72.
25. Thomson, *To the Central African Lakes*, 1:38.
26. Rockell, "Decentering Exploration."
27. Burton, *Lake Regions*, 112
28. Rockel, *Carriers of Culture*, 174
29. Rockel, "Caravan Porters of the Nyika," 151.
30. Rockel, *Carriers of Culture*, 179.
31. Sheldon, *Sultan to Sultan*, 121; this quote was brought to my attention by Rockel, "Caravan Porters of the Nyika," 148.

32. James A. Grant, "27 Photographs of Zanzibar," Picture Library, X0073/018808, RGS Archive.
33. Burton, *Lake Regions*, 1:204.
34. Speke, *What Led to the Discovery*, 294, 296.
35. Grant, *Walk across Africa*, 34.
36. See, for instance, Speke, Report from Speke to Secretary of RGS, February 17 1861, JMS 3/53 RGS Archives, fol. 34.
37. Grant, *Walk across Africa*, 212.
38. Bourke, *Story of Pain*, 192–206.
39. Grant, *Walk across Africa*, 23.
40. Burton, *Lake Regions*, 1:176, 215, 219, 222, 391, 405; 2:131, 157, 162, 229, 236; Speke, *Journal of the Discovery*, 146, 401, 407; Speke, *What Led to the Discovery*, 252.
41. Richard Burton to Norton Shaw, June 24, 1858, RGS/CB/4/292, RGS Archives, fol. 2.
42. Burton, *Lake Regions*, 2:162–63.
43. Grant, *Walk across Africa*, 189–90.
44. Livingstone, *Last Journals*, 2:268–69.
45. David Livingstone, "Field Diary XVI," December 1, 1872–April 6, 1873, Livingstone Online, https://livingstoneonline.org/in-his-own-words/catalogue?query=NOT+letter+NOT+spectral+NOT+shire+NOT+rumor+NOT+blotting+NOT+signatures+NOT+extract+NOT+liv_000049&access=both_transcription&f%5B%5D=dateRangeYear_mi%3A%5B1866+TO+1873%5D&view_pid=liv%3A000016&view_page=0, 65.
46. Livingstone, *Last Journals*, 2:269.
47. Schoenbauer Thurin, *Victorian Travelers*, 144–45. More broadly, see Birkett, *Spinster Abroad*; and Blunt, *Travel, Gender, and Imperialism*.
48. Bishop, *Yangtze Valley*, 118, 105.
49. Bishop, *Yangtze Valley*, 159.
50. Bishop, *Yangtze Valley*, 148, 146.
51. Bishop, *Yangtze Valley*, 137.
52. Stoddart, *Life of Isabella Bird*, 313.
53. Bishop, *Yangtze Valley*, 196, 202, 296.
54. Bishop, *Yangtze Valley*, 202.
55. Alp, "Dogs of the British Antarctic Expedition"; May and Lewis, "Strict Injunctions"; Jones, *Last Great Quest*.
56. Garrard, *Worst Journey in the World*, 2:498.
57. Scott, *Voyage*, 2:491.
58. Scott, *Journals*, 362; these comments are echoed by people who had tried to reenact Scott's expedition. Rees, *Blizzard*, 162–63.

59. Scott, *Journals*, 362.
60. Jones, *Last Great Quest*, 117–20.
61. Scott, *Voyage*; Scott, *Journals*.
62. Armston-Sheret, "Tainted Bodies."
63. Jones, *Last Great Quest*.
64. Markham, "Robert Falcon Scott," 217.
65. Patterson, "Scott's Last Expedition," *Bookman*, December 1913, 66–67.
66. Jones, *Last Great Quest*, 112–14.
67. Twigg, "Carework as a Form of Bodywork," 405; see also McDowell, *Working Bodies*, 169.
68. McDowell, *Gender, Identity, and Place*; Oakley, *Sociology of Housework*.
69. Isabella L. Bird to John Murray, April 11, 1890, MS 42027, NLS, fol. 5.
70. Bishop, *Journeys in Persia*, 1:210.
71. Bishop, *Journeys in Persia*, 1:341.
72. Bishop, *Yangtze Valley*, 207.
73. Bird Bishop, *Korea and Her*, 1:69, 91.
74. Bird Bishop, *Among the Tibetans*, 34.
75. Bishop, *Journeys in Persia*, 1:338–39.
76. Quoted in Bird, *Letters to Henrietta*, 189.
77. Morin, *Frontiers of Femininity*, 54; Bird, *Lady's Life*, 258, 244.
78. Adair, "Feringhi Hakīm."
79. Bishop, *Yangtze Valley*, 420.
80. Bates, *Naturalist on the River Amazons*, 1:137.
81. Burton, *Lake Regions*, 1:279.
82. Burton, *Lake Regions*, 1:132, 2:202.
83. Burton, *Lake Regions*, 1:132.
84. Burton, *Lake Regions*, 1:279, 2:234.
85. Speke, *Journal of the Discovery*, 145.
86. Speke, *What Led to the Discovery*, 288–89.
87. Stanley, *How I Found Livingstone*, 566, 424.
88. Grant, *Walk across Africa*, 22, 78, 75, 113; Speke, *Journal of the Discovery*, 555
89. Rockel, *Carriers of Culture*; see also Rockel, "Decentering Exploration."
90. Speke, *Journal of the Discovery*, 466.
91. Grant, *Walk across Africa*, 360.
92. Thomson, *To the Central African*, 1:95.
93. Rockel, "Enterprising Partners," 757.
94. Burton, *Lake Regions*, 2:161–62.
95. Rockel, "Enterprising Partners," 758, 750.
96. Burton, *Lake Regions*, 2:161–62.
97. Grant, *Walk across Africa*, 43.

98. There were other traditions; for instance, expeditions by men like John Rae drew more on the skills of Indigenous people.
99. Maddison, *Class and Colonialism*; Anthony, *Hoosh*.
100. Scott, *Journals*, 228.
101. Oreskes, "Objectivity or Heroism?," 87–113.
102. Shackleton, *Heart of the Antarctic*, 1:210–15.
103. Strange, "Reconsidering the 'Tragic' Scott Expedition," 74.
104. Priestley, *Antarctic Adventure*, 27–28; see also Evans, "Terra Incognita."
105. Priestley, *Antarctic Adventure*, 66; Scott, *Voyage*, 1:357.
106. Scott, *Voyage*, 1:357.
107. Priestley, *Antarctic Adventure*, 75.
108. Strange, "Reconsidering the 'Tragic' Scott Expedition," 71.
109. Scott, *Voyage*; Cherry-Garrard, *Worst Journey*.
110. Wheeler, *Cherry*, 61
111. For example, Cherry-Garrard, *Worst Journey*, 2:330.
112. Scott, *Voyage*, 1:442–46, 458, 476–78.
113. Scott, *Voyage*, 1:477.
114. For example, Evans, *South with Scott*.
115. Wylie, "Becoming Icy," 259; Strange, "Reconsidering the 'Tragic' Scott Expedition."
116. Edward Wilson, "Journal November 1, 1911 to February 27, 1912," MS 1043;BJ, SPRI, entries December 15, 1911, December 17, 1911, January 7, 1912, January 20, 1912, January 24–28, 1912.
117. Wylie, "Becoming Icy," 259.
118. Shackleton, *Heart of the Antarctic*, 1:342.
119. Wilson, "Medical Aspect," 78.
120. Scott, *Voyage*, 1:483.
121. Driver, "Hidden Histories Made Visible?"
122. Wisnicki, *Fieldwork of Empire*.
123. Bressey, "Surfacing Black and Brown."
124. Livingstone, "'Body' of Evidence."
125. Waller, "Introduction," in *Last Journals*, 1:iv and front matter; similarly, see Stanley, *How I Found Livingstone*, 465–66.
126. Gapah, *Out of Darkness Shining Light* offers a fictional retelling of this scene; this book was brought to my attention by Lioba Hirsch at the 2020 RGS-IBG conference. See also Jeal, *Livingstone*, 378.

3. DISCIPLINED BODIES

1. Peary quoted in Weems, *Peary*, 71.
2. Stanley, *How I Found Livingstone*, 70

3. Driver, "Intermediaries and the Archives," 12.
4. For example, Augustine, *Confessions.*
5. Foucault, *Discipline and Punish*, 128, 136; see also Forth, *Masculinity in the Modern West*, 4–6.
6. Driver, "Power, Space, and the Body," 427; Foucault, *Discipline and Punish*, 135–70.
7. Burton, *Supplementary Papers*, xxviii.
8. Galton, *Art of Travel*, 79, 80.
9. John Hanning Speke to RGS, January 25, 1860, RGS/JHS/1/23a, RGS Archives, fol. 7.
10. Thomson, *To the Central African Lakes*, 1:123.
11. *Report of the Admiralty Committee*, vii, 126, 213.
12. Sir Clements Markham, "Considerations Respecting the Choice of a Leader of an Antarctic Expedition," 1898, National Antarctic Expedition Collection, AA/3/2/1, RGS Archives, fol. 1.
13. Markham, "Considerations," RGS, fol. 1.
14. Newman, *Paths without Glory*, 101–2.
15. Burton *Lake Regions*, 1:151; Burton, *Zanzibar*, 2:256. On medical thinking about the sun in this period, see Kennedy, "Perils of the Midday Sun."
16. Richard Burton to Norton Shaw, June 24, 1858, RGS/CB4/292, RGS Archives, fol. 2.
17. Burton *Lake Regions*, 1:84.
18. Stanley, *How I Found Livingstone*, 277.
19. Thomson, *To the Central African Lakes*, 1:279.
20. Fabian, *Out of Their Minds*, 3.
21. Scott, *Voyage*, 2:67, 83.
22. Robert F. Scott, "Sledging Journal, November 2, 1902 to February 9, 1903," MS 1464/3; BJ, SPRI, entries January 28 and 29, 1903.
23. Speke, *Journal of the Discovery*, xxvii, xxx.
24. Quote in Speke, *Journal of the Discovery*, xxx; for portrayals of African indiscipline, see 187, 190, 216–17.
25. Thomson, *To the Central African Lakes*, 1:289–90.
26. Kingsley, *Travels in West Africa*, 589, 591, 604, 580.
27. Scott, *Voyage*, 2:256–67.
28. Scott, *Voyage*, 1:335.
29. Bird, *Lady's Life*, 67.
30. Bird, *Lady's Life*, 67–69, 71.
31. Speke, *Journal of the Discovery*, 11.
32. Rockel, "Slavery and Freedom."

33. John Hanning Speke to Colonel Rigby, December 12, 1860, RGS/JHS/1/41, RGS Archives.
34. Stanley, *How I Found Livingstone*, 292.
35. Speke, *Journal of the Discovery*, xxvii.
36. Speke, *Journal of the Discovery*, 432–33, 592; Speke to secretary of the RGS, February 17, 1861, JMS3/53, RGS Archives, fol. 34.
37. Speke, *Journal of the Discovery*, 66; Stanley, *How I Found Livingstone*, 337–38.
38. Speke, *Journal of the Discovery*, 513.
39. Speke to secretary of the RGS, February 17, 1861, JMS3/53, RGS Archives, fols. 43, 24. When he returned to east Africa in 1860, Speke was surprised by how difficult it was to recruit porters due to political changes in Unyanyembe.
40. Stanley, *How I Found Livingstone*, 48.
41. Quote Burton, *Lake Regions*, 1:144; see also Rockel, *Carriers of Culture*, 70.
42. Rockel, *Carriers of Culture*, 70; see also Burton, *Lake Regions*, 2:338–39.
43. Rockel, *Carriers of Culture*, 73.
44. Thomson, *To the Central African Lakes*, 1:84; Burton, *Lake Regions*, 1:346–47. Discussion of their dress quoted from Thomson; description of the flag in Burton's account.
45. Thomson, *To the Central African Lakes*, 1:84–85.
46. Burton, *Lake Regions*, 1:144.
47. Thomson, *To the Central African Lakes*, 1:85.
48. E.g., Stanley, *How I Found Livingstone*, 75, 181–82.
49. Thomson, *To the Central African Lakes*, 1:31, 32, 35.
50. Thomson, *To the Central African Lakes*, 1:32–33.
51. Bird, *Unbeaten Tracks* 1:51.
52. Bird, *Unbeaten Tracks*, 2:318.
53. Isabella Bird to John Murray, August 11, 1878, NLS, MS.42024, fol. 104.
54. Bird, *Unbeaten Tracks*, 1:144.
55. Elliot, "Ito and Isabella."
56. Bird, *Unbeaten Tracks*, 1:311, 156.
57. Bird, *Unbeaten Tracks*, 2:137–38.
58. Bird, *Unbeaten Tracks*, 1:156; for a similar analysis, see Elliot, "Ito and Isabella."
59. Elliot, "Ito and Isabella."
60. Kim, "River Control," 666–68.
61. Schoenbauer Thurin, *Victorian Travelers*, 143.
62. Bishop, *Yangtze Valley*, 124.
63. Bishop, *Yangtze Valley*, 117; Isabella Bird to John Scott Keltie, October 5, 1894, RGS/CB, RGS Archives, fol. 5.
64. Bates, *Naturalist on the River*, 1:219.

65. Bates, *Naturalist on the River*, 1:149.
66. "The Equipment and Fitting Out of the Arctic Expedition of 1875," in *Report of the Admiralty Arctic Committee*, (1875), 5, in the Albert Markham Collection, MRK 45/3, National Maritime Museum Archives.
67. Nares, in "Minutes of Evidence," *Report of the Admiralty Committee on Scurvy*, 1.
68. Markham, *Great Frozen Sea*, 5.
69. James Reid to Anne Reid, May 13, 1845, in *May We Be Spared*, 104.
70. "Service in the Antarctic Expedition," 1901, RGS/AA/4/1, RGS Archives, fol. 1.
71. See file "Crew," RGS/AA/3/4/1–48, Antarctic Expedition Collection, RGS Archives.
72. "Crew," RGS/AA/3/4/1–48, Antarctic Expedition Collection, RGS Archives; Armitage, *Two Years*, 8.
73. Armitage, *Two Years*, 121.
74. Mossakowski, "Sailors Dearly Love to Make Up," 200. On working-class fitness culture, see also Scott, "Body-Building and Empire-Building"; and Begiato, *Manliness in Britain*, 196.
75. "Report of Medical Examination of Officer of Expedition," January 7, 1901, RGS/AA 3/1/5, Antarctic Expedition Collection, RGS Archives; Medical Department Admiralty to Secretary of the National Antarctic Expedition, "Medical Report on Scientific Members of the Expedition," July 10, 1901, RGS/AA/3/1/18, Antarctic Expedition Collection RGS Archives; "Average Consumption of Alcohol Beverages," n.d., RGS/AA/3/2/11, Antarctic Expedition Collection, RGS Archives.
76. Comments in Clements Markham's hand on letter from Medical Department Admiralty, July 10, 1901, RGS/AA/3/1/18, RGS Archives.
77. Robert F. Scott to Reginald Koettlitz, February 19, 1903, MS 366/14; ER/24, SPRI.
78. Seligmann, *Rum, Sodomy, Prayers*, 43, 82.
79. Johnston, "Arbitrary and Cruel Punishments."
80. Sears, "Discipline in the Royal Navy."
81. Seligmann, *Rum, Sodomy, Prayers*, 96–97, 138–39; Sears, "Discipline in the Royal Navy," 39–60.
82. Sir John Franklin to John Adam, May 19, 1845, in *May We Be Spared*, 128.
83. James Fitzjames to Elizabeth Conningham, June 9, 1845, in *May We Be Spared*, 190–91.
84. Markham, *Great Frozen Sea*, 144, 313.
85. Scott, *Voyage*, 1:75.
86. Scott, *Voyage*, 1:292–307, 75.

87. Strange, "Reconsidering the 'Tragic' Scott Expedition," 80.
88. Scott, *Journals*, 92.
89. Cherry-Garrard, *Worst Journey*, 2:547.
90. Evans, *South with Scott*, 116.
91. Conley, *From Jack Tar to Union Jack*.
92. Scott, *Journals*, 469.
93. Priestley, *Antarctic Adventure*, 80.
94. Priestley, *Antarctic Adventure*, 80. The two men were not cooks before the expedition but took on this role; for a broader history of the party, see Hooper, *Longest Winter*.
95. Priestley, *Antarctic Adventure*, 80.
96. Naylor, "Log Books and the Law of Storms."
97. Driver, *Geography Militant*.
98. Stanley, *How I Found Livingstone*, 317, 712, 108, 140.
99. David Livingstone, "Field Diary IV, July 1–September 5, 1866," July 7, 1866, fol. 18, Livingstone Online, https://livingstoneonline.org/in-his-own-words/catalogue?query=NOT+letter+NOT+spectral+NOT+shire+NOT+rumor+NOT+blotting+NOT+signatures+NOT+extract+NOT+liv_000049&access=both_transcription&f%5B%5D=dateRangeYear_mi%3A%5B1866+TO+1873%5D&view_pid=liv%3A000004&view_page=0, 18.
100. Thomson, *To the Central African Lakes*, 1:280.
101. Quote in Burton, *Lake Regions*, 1:361–62; Newman, *Paths without Glory*, 127. Newman analyses how likely it is that Burton was an authoritarian commander.
102. Speke, *Journal of the Discovery*, 39.
103. Cameron, *Across Africa*, 1: 07.
104. Grant, *Walk across Africa*, 32.
105. Speke, *Journal of the Discovery*, 164.
106. Speke to Secretary of the RGS, February 17, 1861, JMS 3/53, fol. 36.
107. Speke, *Journal of the Discovery*, 271; Stanley, *How I Found Livingstone*, 28. Stanley noted the visibility of the scar when he hired Bombay for the Livingstone Search Expedition.
108. Burton, *Lake Regions*, 2:239; see also Galton, *Art of Travel*, Richard Francis Burton Library Manuscripts Collection, RFB 381, Huntington Library, 84–86. Burton's comments on the latter also demonstrate his support for violent punishments.
109. Wadiwel, "Thick Hides," 54, 50. More broadly, see Harris, *Punishing the Black Body*; and Pierce and Anupama, "Discipline and the Other Body."
110. Speke, *Journal of the Discovery*, xxx; Richard Burton to Norton Shaw, June 24, 1858, RGS/CB4/292, RGS Archives, fol. 2.

111. Speke, *Journal of the Discovery*, xxx; Burton, *Lake Regions*, 1:142, reports similar responses.
112. Kingsley, *Travels in West Africa*, 579.
113. Kearns, *Geopolitics and Empire*, 109–11.
114. Stanley, "Field Notebook," RMCA, n.d. 1888; Bonny, "Autograph Journal," RMCA, February 9, 1888.
115. Bonny, "Autograph Journal," RMCA, February 9, 1888.
116. Stanley, "Field Notebook," RMCA, n.d. 1888.
117. "Dark Deeds in Darkest Africa," *Pall Mall Gazette*, November 10, 1890, 6. The exact circumstances and truthfulness of these allegations is a source of debate.
118. Bonny, "Autograph Journal," April 23–25, 1888; see also Barttelot, *Life*, 228–29.
119. Stanley, "Field Notebook," RMCA, n.d. 1888; Bonny, "Autograph Journal," RMCA, July 19, 1888, July 26, 1888.
120. "The Emin Relief Expedition," *Annual Report of the British & Foreign Aborigines' Protection Society*, December 1, 1890.
121. John Rose Troup, another of the expedition's officers, quoted in "The African Scandals," *Lloyd's Illustrated Newspaper*, November 16, 1890.
122. Anthony, *Hoosh*, 20.
123. Fiennes, *Shackleton*, 295.

4. ANIMAL BODIES

1. Oates, *Matabeleland*, 265, 262.
2. Speke to Secretary of the RGS, February 17, 1861, JMS 3/53, RGS, fol. 41.
3. Forsyth, "More-than-Human Geographies," 530.
4. Forsyth, "More-than-Human Geographies"; Jones, "Animals"; Kennedy, *Last Blank Spaces*, 142–46; Leane, "Animals"; Murray, "Use and Abuse of Dogs"; Oliver, "Animals"; Oliver, "Labour, Collaboration, Conflict"; Wilks, "Coldest Dog and Pony Show."
5. Whatmore, *Hybrid Geographies*; Graham Burnett, "It Is Impossible." The latter also analyzes expeditions as hybrid undertakings, though focusing only on humans.
6. For example, Haraway, *When Species Meet*; Derrida, *Animal That Therefore I Am*.
7. For example, Morin, "Carceral Spaces"; Morin, *Carceral Space*; Oliver, *Veganism, Archives, and Animals*; Oliver, "Beyond-Human Research"; Philo and Wilbert, *Animal Spaces*; Urbanik, *Placing Animal*; and Whatmore, *Hybrid Geographies*.

8. Sivasundaram, "Imperial Transgressions," 169. On animals and empire more broadly, see Saha, "Colonizing Elephants"; Skabelund, "Animals and Imperialism"; Cribb et al., *Wild Man from Borneo*; Saha, *Colonizing Animals*; Burton and Mawani, *Animalia*; and Skabelund, "Animals and Imperialism."
9. Hribal, "Animals Are Part of the Working Class."
10. Boisseron, *Afro-Dog*; Iman Jackson, *Becoming Human*; McRae, "Great Chain of Being"; Morin, "Carceral Spaces."
11. For example, Adams, *Sexual Politics of Meat*; and Boisseron, *Afro-Dog*.
12. Sivasundaram, "Imperial Transgressions," 156; see also Sivasundaram, "Prehistory of COVID-19."
13. Anderson and Perrin "How Race Became Everything"; Qureshi, *Peoples on Parade*, 200; Robinson, *Lost White Tribe*, 90–91.
14. Scott, *Voyage*, 2:13; see also 1:327, 472.
15. Scott *Voyage*, 1:472.
16. Shackleton, *Heart of the Antarctic* 1:37; Scott, *Journals*, 265.
17. Johnston, "Beyond the Clearing," 634; for discussion of these debates in relation to history, see Swart, *Riding High*, 1–17.
18. Jones, "(Un)ethical Geographies," 283.
19. Swart, "Introduction," 97.
20. Livingstone, *Missionary Travels*, 12–13.
21. Kingsley, *Travels in West Africa*, 88.
22. Burton, *First Footsteps in East Africa*, 1:17n2. This quote was brought to my attention by Sadia Aidid on Twitter, March 25, 2015 (account deleted).
23. Dobson, "Medical Hints," 253.
24. David Livingstone to Robert Gray, April 8, 1860, Brenthurst Library, fol. 4, Livingstone Online, https://livingstoneonline.org/in-his-own-words/catalogue/?f%5B0%5D=repository_ms%3A%22Brenthurst+Library%22&view_pid=liv%3A001560&view_page=0.
25. Burton, *Lake Regions*, 1:156.
26. Kingsley, *Travels in West Africa*, 92.
27. Speke, *What Led to the Discovery*, 224–25.
28. Thomson, *To the Central African Lakes*, 1:130–31, 90.
29. Bates, *Naturalist on the River*, 2:290.
30. Thomson, *To the Central African Lakes*, 1:130–31, 90.
31. Stanley, *How I Found Livingstone*, 338.
32. Mackenzie, *Empire of Nature*, 22.
33. Mackenzie, *Empire of Nature*, 305.
34. Montefiore, "Jackson-Harmsworth North Polar Expedition," 511.
35. Jackson, *Thousand Days*, 1:171.
36. Bates, *Naturalist on the River*, 2:260.

37. Bates, *Naturalist on the River*, 2:259–62.
38. Ook-barloo in Nourse, *Narrative of the Second Arctic*, 590–91; this was brought to my attention by Woodman, *Unravelling the Franklin Mystery*, 292.
39. Woodman, *Unravelling the Franklin Mystery*, 291–92, 228.
40. Speke, *Journal of the Discovery*, 619.
41. James A. Grant to Richard Blanshard, March 8, 1861 (copy), JMS/3/53, RGS Archives, fols. 3–4.
42. Speke, *What Led to the Discovery*, 173.
43. Samuel Baker, "The First Rush of the Giraffes / Bagging the Giraffes at a Long Shot about 100 Yards, Sofi," Picture Library, S0026488; and Baker, "Close Quarters with the Lion," Picture Library, S0025495, RGS Archives.
44. Baker, *Rifle and the Hound*, v.
45. John Hanning Speke to Secretary of the RGS, February 17, 1861, JMS/3/53, RGS Archives, fol. 34.
46. Grant, *Walk across Africa*, 32.
47. Stanley, *How I Found Livingstone*, 594.
48. Grant, *Walk across Africa*, 32–34.
49. Mackenzie, *Empire of Nature*, 298.
50. John Hanning Speke to Secretary of the RGS, February 17, 1861, JMS/3/53, RGS Archives, fol. 34.
51. Stanley, *How I Found Livingstone*, 337.
52. Stanley, *How I Found Livingstone*, 342.
53. Speke, *Journal of the Discovery*, 66.
54. Mrinalini, *Colonial Masculinity*.
55. Scott, *Voyage*, 1:149.
56. Scott, *Journals*, 73–74. Ponting's own account of the incident can be found in Ponting, *Great White South*, 64–65. This incident is a rare one, as there have been very few deliberate Orca attacks on humans.
57. Scott, *Journals*, 73–74.
58. Ponting, *Great White South*, 65–66.
59. Scott, *Journals*, 74–75.
60. Darwin, *Zoology*.
61. Wulf, *Invention of Nature*.
62. Dritsas, "Lake Nyassa to Philadelphia"; more broadly, see MacGregor, *Naturalists in the Field*.
63. On zoology, see Sylph, "Women in Zoology"; on geography, see Bell and McEwan, "Admission of Women Fellows."
64. Kingsley, *Travels in West Africa*, 7.
65. Morris, "Reflections on Some Practical Aspects."
66. McCook, "It May be Truth."

67. Burton, *Two Trips*, 1:112.
68. Burton, *Two Trips*, 1:39.
69. Burton, *Two Trips*, 1:113.
70. Burton, *Two Trips*, 1:174, 175, 176.
71. Bates, "Natural History," 203; see also Driver, *Geography Militant*.
72. Bates, *Naturalist on the River*, 2:162–63, 309.
73. Jackson, *Thousand Days*, 1:165, 175, 219; Montefiore, "Jackson-Harmsworth North Polar Expedition," 513.
74. Frank Oates, *Matabeleland*, 131–32.
75. McRae, "Great Chain of Being."
76. Bates, *Naturalist on the River*, 2:238, 236.
77. Oliver, "Labour, Collaboration, Conflict, and Discovery."
78. Nagai, "Vermin Writing"; Skotnes-Brown, "Scurrying Seafarers."
79. David, "Elegant Amazons."
80. Isabella Bird to John Murray, April 11, 1890, MS 42027, NLS, fol. 6.
81. Uchiyama et al., "Three-Dimensional Analysis."
82. Haraway, *Companion Species Manifesto*.
83. Bird, *Lady's Life*, 156, 181.
84. Bishop, *Among the Tibetans*, 9.
85. Swart, *Riding High*; Worboys et al., *Invention of the Modern Dog*.
86. Bird, *Unbeaten Tracks*, 1:314.
87. Bird Bishop, *Korea and Her Neighbours*, 1:137–51.
88. Bird Bishop, *Korea and Her Neighbours*, 1:139.
89. Grant, *Walk across Africa*, 23. For other discussions of these animals, see also Burton, *Lake Regions* 1:10–12, 46; John Hanning Speke to George Grey, November 1, 1860, George Grey Collection, GL S 39.3, Auckland Public Libraries; Stanley, *How I Found Livingstone*, 68; Burton, *Supplementary Papers to the Mwata*, xxv; and Thomson, *To the Central African Lakes*, 1:120–21.
90. Simpson, *Dark Companions*, 17; Kennedy, *Last Blank Spaces*, 147.
91. Grant, *Walk across Africa*, 28.
92. Speke to Secretary of the RGS (large text copy), February 17, 1861, JMS 3/53, RGS Archives, fols. 44–45.
93. Burton, *Lake Regions*, 1:64; Grant, *Walk across Africa*, 28.
94. Quote Speke, *Journal of the Discovery*, 165; see also Grant, *Walk across Africa*, 117.
95. Thomson, *To the Lakes of Central Africa*, 121.
96. Jackson, *Great Frozen Land*, 70. For superiority of reindeer to horses, see 123; for hopes to use them on Arctic expedition, see 166.
97. Jackson, *Great Frozen Land*, 181–82.
98. Shackleton, *Heart of the Antarctic*, 1:21.

99. Joseph James Kinsey to Mr. C. Holdsworth, Joseph James Kinsey Collection, Alexander Turnbull Library: National Library of New Zealand, May 15, 1911, fols. 1–2.
100. Riffenburgh, *Shackleton's Forgotten Expedition*, 119–20.
101. Cherry-Garrard, *Worst Journey*, 1:114.
102. Scott, *Journals*, 15–16, 85, 124, 118, 138, 154, 228, 318; Cherry-Garrard, *Worst Journey*, 1:218–24.
103. Cherry-Garrard, *Worst Journey*, 1:219.
104. Scott, *Journals*, 338.
105. Barczewski, *Antarctic Destinies*, 50.
106. Scott, *Voyage*, 1:464.
107. Huntford, *Last Place on Earth*, 410–507.
108. Albert Armitage to Clements Markham, June 5, 1900, AA7/1/1, AA Collection, RGS Archives, fol. 8.
109. Folder AA/7/1–6, RGS Archives, discusses the purchase of these dogs. Scott, *Voyage*, 1:471.
110. Scott, *Voyage*, 1:463–73; Jones, *Last Great Quest*, 66.
111. Scott, "Brief Summary of Proceedings at 'Discovery' Winter Quarters," February 23, 1903, Royal Society, MS/591/1/5, fol. 25. See also Scott, *Voyage*, 2:103–4; and Robert Falcon Scott, "Scott Sledging Journal Summer 1902," SPRI MS1464/4, SPRI.
112. Scott, *Voyage*, 1:467.
113. Forth, *Masculinity in the Modern West*, 160–64.
114. Scott, *Voyage*, 1:465–67; Jones, *Last Great Quest*, 118; Wilks, "Coldest Dog and Pony Show."
115. Heilprin, "British National Antarctic Expedition," 179.
116. Amundsen, *South Pole*, 28.
117. Gregory, *Of Victorians and Vegetarians*.
118. Alp, "Dogs of the British Antarctic Expedition," 479.
119. Scott, *Voyage*, 1:256–57.
120. Scott, *Voyage*, 1:256–57.
121. Scott, *Voyage*, 2:13.
122. Scott, *Journals*, 116.
123. Bishop, *Among the Tibetans*, 62, 98, 63.
124. Qiu et al., "Yak Genome."
125. Bishop, *Among the Tibetans*, 63.
126. Phoenix "More like a Picnic Party," 248; Kennedy, *Last Blank Spaces*, 142.
127. Kennedy, *Last Blank Spaces*, 142.
128. Phoenix "More like a Picnic Party," 259–73; the camels purchased in Yemen were referred to as "Egyptian" by the explorers.

129. Phoenix "More like a Picnic Party," 245.
130. Phoenix "More like a Picnic Party," 305–8.
131. Kennedy, *Last Blank Spaces*, 146.
132. For complaints, see Lindsay, *Journal of the Elder*, 22, 33, 35, 42, 46; for descriptions of their work, see 21, 53, 57.
133. Lindsay, *Journal of the Elder*, 26
134. Kennedy, *Last Blank Spaces*, 145.
135. Joseph James Kinsey to Robert Falcon Scott, August 20, 1910, Joseph James Kinsey Collection, Alexander Turnball Library, National Library of New Zealand, fol. 4; Cherry-Garrard, *Worst Journey*, 1:224.
136. Cherry-Garrard, *Worst Journey*, 1:224.
137. Scott, *Journals*, 95; Cherry-Garrard, *Worst Journey*, 1:224.
138. Scott, *Journals*, 73, 85, 95, 170, 177, 187.
139. Bird Bishop, *Korea and Her Neighbours*, 1:138–39.

5. SENSING BODIES

1. Arnold, *Tropics*, 29; Said, *Orientalism*, 214; Ryan, *Cartographic Eye*; Smith, *European Vision*. There is a vast literature on "the gaze"; see Berger, *Ways of Seeing*, Foucault, *Discipline and Punish*; Said, *Orientalism*; Otter, *Victorian Eye*; Foucault, *Birth of the Clinic*; and bell hooks, *Black Looks*.
2. Arnold, *Tropics*, 31.
3. Rose, "On the Need to Ask," 217; she draws on Pratt, *Imperial Eyes*, 9. See also Rose, *Feminism and Geography*.
4. For example, Clayton, *Islands of Truth*; Coen, *Climate in Motion*; Fabian, *Out of Our Minds*; Fleetwood, "No Former Travellers"; Fleetwood, *Science on the Roof*; *Kennedy, Last Blank Spaces*; Lamb, *Scurvy*; Mueggler, *Paper Road*; and Wylie, "Becoming Icy."
5. Freshfield, "Introduction," in *Hints to Travellers*, 7th ed., 5. *Hints to Travellers* was first published in 1851 and went through multiple editions, being expanded greatly in 1883. This advice was first included in the 1893 edition.
6. Stuart, *Explorations in Australia*, 20.
7. In this sense I have been influenced by Carter, *Road to Botany Bay*. On oral information, see Lambert, *Mastering the Niger*; Wisnicki, *Fieldwork of Empire*; and Withers, "Mapping the Niger."
8. Jay, *Downcast Eyes*, 6. See also 22; and Rodaway, *Sensuous Geographies*, 27. On distance and objectivity, see Fabian, *Out of Our Minds*, 185–86.
9. Fabian, *Out of Our Minds*, 185–86.
10. Sullivan, *Geography of the Everyday*, 144; see also Paterson, "Haptic Geographies."

11. Sullivan, *Geography of the Everyday*, 144; Merleau-Ponty, *Phenomenology of Perception*.
12. Wylie, "Becoming Icy"; Crang, "Qualitative Methods?" On touch historically, see Classen, *Deepest Sense*.
13. Classen, "Introduction," 23.
14. Otter, *Victorian Eye*.
15. Livingstone, *Geographical Tradition*, 98; see also Cosgrove, *Geography and Vision*.
16. Classen, "Introduction," 6.
17. Otter, *Victorian Eye*, 254.
18. Classen, "Introduction," 6.
19. Godlewska, "From Enlightenment Vision," 236.
20. Leake, "Is the Quorra," 3; for an analysis of this quote, see Barnett, "Impure and Worldly Geography," 244.
21. Dettelbach, "Stimulations of Travel," 43–58.
22. Driver, "Material Memories of Travel," 33.
23. Driver, "Distance and Disturbance," 76; here he draws on Dettelbach, "Global Physics and Aesthetics Empire." On this issue more broadly, see Driver, *Geography Militant*; Lambert, *Mastering the Niger*; McCook, "It May Be Truth"; and Lawrence and Shapin, *Science Incarnate*.
24. Wisnicki, "Cartographical Quandaries"; Wisnicki, "Charting the Frontier"; Wisnicki, *Fieldwork of Empire*; Dritsas, "Expeditionary Science"; Withers, "Mapping the Niger"; Lambert, *Mastering the Niger*; Driver and Martins, *Tropical Visions*.
25. Herschel, *Manual of Scientific Enquiry*, 128.
26. On instruments, see Withers, "Geography and 'Thing Knowledge'"; Wess and Withers, "Instrument Provision and Geographical Science"; and MacDonald and Withers, *Geography Technology and Instruments*. On photography, see Ryan, *Photography and Exploration*.
27. Kennedy, *Last Blank Spaces*, 28.
28. Graham Burnett, *Masters of All They Surveyed*, 111.
29. Ryan, *Photography and Exploration*, 33; Forsyth, "More-than-Human Geographies," 533.
30. "Dr. Cook Does Not Know if He Reached the Pole," *Herald*, January 6, 1911, 5.
31. Caldwell, "Framing Digital Image Credibility," 58.
32. Raj, *Relocating Modern Science*, 188; see also Raj, "When Human Travellers Become Instruments."
33. Raj, *Relocating Modern Science*, 203.

34. John Hanning Speke to Norton Shaw, "Explanation to the Book," July 2, 1858, RGS/JHS/1/2, RGS Archives, fol. 2.
35. Burton, *Lake Regions*, 1:189.
36. Kennedy, *Last Blank Spaces*, 40.
37. Stern, "Exploration and Enlightenment," 65.
38. Livingstone, *Putting Science in Its Place*, 152.
39. Salisbury and Shail, "Introduction," in *Neurology and Modernity*, 21.
40. Bourke, *Story of Pain*, 195–96; Salisbury and Shail, "Introduction," 31.
41. Galton, *Inquiries into Human Faculty*, 25.
42. Bourke, *Story of Pain*, 209.
43. Eastlake, "Lady Travellers." Such advice did affect how some wrote about their travels; see Park, "Land of the Morning Calm."
44. Birkett, *Spinsters Abroad*, 191.
45. Burton, *Lake Regions*, 2:154.
46. Pratt, *Imperial Eyes*, 197–204.
47. Burton, *Lake Regions*, 2:42.
48. Speke, *What Led to the Discovery*, 203.
49. Pratt, *Imperial Eyes*, 201.
50. These comments do not appear in Burton's more scientifically styled report of the expedition for the RGS; Burton, *Lake Regions*, 212. Speke reported that, at the time, Burton the pair were "disabled by sickness—Captain Burton utterly"; Speke, "Journal of a Cruise," 343.
51. Speke, *What Led to the Discovery*, 305–58.
52. Speke, "Captain Speke's Discovery," 412; Speke, *What Led to the Discovery*, 310.
53. Speke, *What Led to the Discovery*, 311; my emphasis.
54. Burton, *Lake Regions*, 2:207.
55. Burton, *Lake Regions*, 2:169. He includes a similar claim in Burton to Shaw, June 24, 1858, RGS/CB, RGS Archives, fol. 2.
56. Burton, *Lake Regions*, 1 84.
57. Hussey, "Imperial Mobilities," 129–71; Hussey, *Imperial Bodies*.
58. Anon., "The Life of Captain Sir Richard F. Burton," *Edinburgh Review*, October 1893, 459.
59. Burton, *Lake Regions*, 1:84.
60. John Hanning Speke to Norton Shaw, July 2, 1858, RGS/JHS/1/2, RGS Archives, fol. 3.
61. Speke to Shaw, July 2, 1858, RGS Archive, fol. 3.
62. Richard Burton to Norton Shaw, June 24, 1858, RGS/CB, RGS Archives, fol. 2.
63. John Hanning Speke to Norton Shaw "Explanation to The Book," July 2, 1858, JHS/1/2, RGS Archives, fol. 2.

64. Speke to Shaw, "Explanation to the Book," RGS Archives, fol. 2.
65. Speke to Shaw, "Explanation to the Book"; Speke's 1857 sketch maps also emphasize plant growth, describing large areas using terms like "thorn jungle" or as "thorn covered plain." John Hanning Speke 1857, "12 Sketch Maps Showing the Route of a Journey by Speke and Burton, to and around Sea Ujiji," Map Room Collection, CU18-AFE-TZA-S507, RGS Archives, fols. 5–6.
66. Driver and Martins, "John Septimus Roe," 158.
67. Wisnicki, "Cartographical Quandaries," 471–75, discusses the problems at length; for an insightful summary of these debates, see also Dritsas, "Expeditionary Science."
68. Wisnicki, *Fieldwork of Empire*, 58; see also Wisnicki, "Charting the Frontier," 121.
69. Wisnicki, *Fieldwork of Empire*, 58.
70. Jones, "Local Knowledge," 165–78.
71. Wisnicki, *Fieldwork of Empire*, 63–65.
72. Finkelstein, "Unravelling Speke," 130.
73. Wisnicki, "Cartographical Quandaries," 469.
74. Jones, "Local Knowledge," 141; see also Cox, "Armchair Geography."
75. Quote Burton, *Lake Regions*, 2:207; see also 171. Burton thought that the hand gestures probably meant "a few miles." Burton's marginal comments in John Hanning Speke, *What Led to the Discovery of the Source of the Nile* (London: 1864), Richard Francis Burton Library Manuscripts Collection, RFB 1680, Huntington Library, 312.
76. McQueen, "Captain Speke's Discovery," 125–26; on MacQueen's use of oral information, see Lambert, *Mastering the Niger*.
77. Bird, *Unbeaten Tracks*, 1:318.
78. Bishop, *Journeys in Persia*, 1:322.
79. Bishop, *Yangtze Valley*, x.
80. Bishop, *Yangtze Valley*, ix.
81. Bishop, *Yangtze Valley*, ix.
82. Forsyth, "More-than-Human Geographies," 533.
83. Isabella L. Bishop to John Scott Keltie, March 10, 1897, RGS/CB, RGS Archives, fol. 6. See also, Gartlan, "Complete Craze," 16; and Ireland, *Isabella Bird*, 18.
84. Gartlan, "Complete Craze," 26.
85. Bird to Keltie, March 10, 1897, RGS Archives, fol. 6; Bishop, *Yangtze Valley*, 156; Ireland, *Isabella Bird*, 18–19, 23.
86. Gartlan, "Complete Craze," 26; Hayes, "Geographical Light."
87. Isabella Bird to John Murray, January 23, 1897, MS 2028, NLS, fol. 29.
88. Ryan, *Photography and Exploration*.

89. Gartlan, "Complete Craze," 16–17. This was not uncommon; see Ryan, *Photography and Exploration*, 78.
90. Stoddart, *Life of Isabella Bird*, 298. Again, this was common; Ryan, *Photography and Exploration*, 119–21.
91. Bell, "Eyeing Samoa," 163.
92. Bishop, *Yangtze Valley*, 195.
93. Bishop, *Yangtze Valley*, 207.
94. Bishop, *Yangtze Valley*, 285. Others disagree with this interpretation; see Chang, *Britain's Chinese Eye*, 169.
95. Kennedy, *Last Blank Places*, 266.
96. Scott, *Voyage*, 1:144.
97. Fox, "Walking in Circles," 20.
98. Fox, "Walking in Circles," 25; McClintock, "Arctic Sledge Travelling."
99. Scott, *Voyage*, 1:144.
100. Wylie, "Becoming Icy," 255.
101. Driver, "Material Memories of Travel," 3.
102. Yusoff, "Climates of Sight," 48–63.
103. Rees, "Polar Mirages," 193; Yusoff, "Climates of Sight," 51–53.
104. Scott, *Voyage*, 1:144, 170–72, 177; Scott, *Journals*, 208, 467.
105. Scott, *Voyage*, 1:144.
106. Scott, *Journals*, 279; my emphasis.
107. Carter, *Road to Botany Bay*, 73.
108. Scott, *Voyage*, 1:179.
109. Yusoff, "Configuring the Field," 72–73, 75.
110. Scott, *Journals*, 208.
111. Scott, *Voyage*, 2:389.
112. Scott, *Voyage*, 2:389.
113. Wylie, *Landscape*, 150.
114. Shackleton, *Heart of the Antarctic*, 1:280.
115. Scott, *Journals*, 119, 266; Cherry-Garrard, *Worst Journey*, 2:477.
116. Markham, *Great Frozen Sea*, 272–73.
117. Scott, *Voyage*, 1:504.
118. Edward Wilson, "Diary of the Terra Nova Expedition," February 24, 1912, MS 1043; BJ, SPRI Archive.
119. William Lashly quoted in Cherry-Garrard, *Worst Journey*, 2:385.
120. Barnes, "Senses in Medicine," 154–55; Baldwin, "How Night Air Became."
121. Barnes, "Senses in Medicine," 146.
122. Harrison, *Climates and Constitutions*, 63.
123. Barnes, "Senses in Medicine," 150.
124. Kingsley, *Travels in West Africa*, 97.

125. Thomson, *To the Central African Lakes*, 1:163.
126. Speke, *What Led to the Discovery*, 184.
127. Stanley, *How I Found Livingstone*, 102.
128. Burton, quoted in *Search for the Source*, 79. See also Burton, *Lake Regions*, 1:62, 92, 127, 184; 2:235, 262.
129. Burton, *Lake Regions*, 1:62.
130. Newman, *Paths without Glory*, 94.
131. Thomson, *To the Central African Lakes*, 1:163.
132. Baker, *Albert Nyanza*, 1:105, 188, 222.
133. Burton, *Lake Regions*, 2:236.
134. Grant, *Walk across Africa*, 26.
135. Burton, *Lake Regions*, 1:232.
136. Livingstone, *Zambesi*, 1:602.
137. Jeal, *Livingstone*, 150; Dritsas, *Zambesi*, 188.
138. Newman, *Paths without Glory*, 97; Kennedy, *Magic Mountains*; Beattie, *Empire and Environmental Anxiety*, 39–71; Arnold, "Envisioning the Tropics," 148.
139. Bates, *Naturalist on the River*, 1:9–10.
140. Cherry-Garrard, *Worst Journey*, 1: 253.
141. Edward Wilson, *Scott's Last Expedition*, 2:58.
142. Cherry-Garrard, *Worst Journey*, 1:253.
143. Yusoff, "Antarctic Exposure."
144. Wylie, *Landscape*, 159.
145. Driver, "Hidden Histories."

6. EXAMINED BODIES

1. Burton, *Personal Narrative of a Pilgrimage*, 1 21.
2. Armston-Sheret, "Leaky Bodies."
3. Stocking, *Victorian Anthropology*, 79.
4. Tylor, "Anthropology," 222. Anthropological advice was expanded in future editions of the handbooks; see Driver, *Geography Militant*, 64.
5. Tylor, "Anthropology," 222, 224.
6. For an example of measurements on living people, see Jackson, *Great Frozen Land*, 61. For examples of skull collecting, see Burton, *Zanzibar*, 2:346; and Jackson, *Thousand Days*, 1:24.
7. Qureshi, *Peoples on Parade*; Kaalund, "Erasure as a Tool," 4.
8. Asad, *Anthropology and the Colonial Encounter*; Stocking, *Victorian Anthropology*; Stocking, *Race, Culture, and Evolution*.
9. Fabian, *Out of Our Minds*, 186.
10. Pratt, *Imperial Eyes*, 75–85.
11. Kennedy, *Last Blank Spaces*, 214; see also Kennedy, *Highly Civilized Man*, 115.

12. Kennedy, *Last Blank Spaces*, 216; see also Clayton, *Islands of Truth.*
13. Merleau-Ponty, *Visible and Invisible.*
14. Mueggler, *Paper Road*, 161.
15. Kennedy, *Highly Civilized Man*, 115–16.
16. Said, *Culture and Imperialism*, xiii.
17. Woodman, *Unravelling the Franklin Mystery*; see chapter 3 in Woodman for a discussion of the challenges of Inuit testimony and the way it was recorded.
18. Nourse, *Narrative of the Second Arctic Expedition*, 591.
19. This anecdote was relayed by Iggiara, Manguq's son, quoted in Rasmussen, "Fifth Thule Expedition," 138.
20. Rasmussen, *Netsilik Eskimos*, 128.
21. Nourse, *Narrative of the Second Arctic Expedition*, 607.
22. Rasmussen, *Netsilik Eskimos*, 131.
23. Nourse, *Narrative of the Second Arctic Expedition*, 595–96.
24. Burton, *Lake Regions*, 1:63.
25. Grant, *Walk across Africa*, 39.
26. Kennedy, *Last Blank Spaces*, 216; Fabian, *Out of Our Minds*, 186.
27. Stanley, *How I Found Livingstone*, 176, 186–87.
28. Cameron, *Across Africa*, 1: 285.
29. Speke, *What Led to the Discovery*, 301–2.
30. Speke, *What Led, to the Discovery*, 284–85; Kennedy, *Highly Civilized Man*, 115.
31. Kennedy, *Highly Civilized Man*, 115–16.
32. Burton, *Lake Regions*, 2:128–29; Kennedy, *Highly Civilized Man*, 115.
33. Kennedy, *Highly Civilized Man*, 115–16, 131.
34. Kennedy, *Highly Civilized Man*, 115.
35. Speke, *What Led to the Discovery*, 277. The use of "mumbled" in the passage is strange and is, perhaps, a misprint; it may also reflect Speke's odd writing style. However, the word can be used to describe lightly chewing something, raising the possibility that Speke's fingers were placed in her mouth.
36. Kennedy, *Last Blank Spaces*, 223.
37. Speke, *What Led to the Discovery*, 276–78.
38. Burton and Speke, "Exploration in East Africa," October 27, 1859, JMS/3/49, RGS Archives, 6.
39. Brierley and Spear, "Mutesa," 601. See also Reid "Ganda on Lake Victoria"; and Robinson, *Lost White Tribe*, 96–97.
40. Brierly and Spear, "Mutesa," 601.
41. Reid, "Ganda on Lake Victoria"; Reid, "Images of an African Ruler," 283; Robinson, *Lost White Tribe*, 97. For another explorer's engagement with Mutesa, see Stanley, *Through the Dark Continent*.

42. Speke, *Journal of the Discovery*, 352, 355.
43. Speke, "First Corrected Proofs," 262.
44. Casada, "James A. Grant," 246.
45. Grant, *Walk across Africa*, 222.
46. Speke, *Journal of the Discovery*, 357.
47. Speke, *Journal of the Discovery*, 363, 357, 372–73.
48. Speke, *Journal of the Discovery*, 357, 408.
49. Speke, *Journal of the Discovery*, 357, 408; Grant, *Walk across Africa*, 223.
50. Brierley and Spear, "Mutesa," 603.
51. Bird, *Unbeaten Tracks*, 1:178.
52. Park, "Land of the Morning Calm," 529, 534.
53. Sims, *Japanese Political History*.
54. Park, "Land of the Morning Calm," 529, 534.
55. Isabella Bird to John Murray, August 11, 1878, MS 42024, NLS, fol. 104.
56. Bird, *Unbeaten Tracks*, 1:178.
57. Bird, *Unbeaten Tracks*, 1:290.
58. Bird, *Unbeaten Tracks*, 1:164–65, 177–78, 182, 258, 270, 300.
59. Bird, *Unbeaten Tracks*, 1:256.
60. Bird, *Unbeaten Tracks*, 1:168, 165.
61. On these observations, see Kata, "I Write the Truth as I See It"; Spiker, "'Civilized' Men and 'Superstitious' Women"; Park, "Missing Link Found"; and Williams, "Like the Ladies of Europe?"
62. Bird, *Unbeaten Tracks*, 1:338.
63. Kanasaka, *Isabella Bird*, 3–7; Elliot, "It's Japan," 1–20.
64. Bird, *Unbeaten Tracks*, 1:311–12.
65. Elliot, "Ito and Isabella"; Bhabha, "Of Mimicry and Man," 125–33; Kennedy, *Last Blank Spaces*, 208.
66. Bird, *Unbeaten Tracks*, 1:156.
67. Elliot, "Ito and Isabella."
68. Kanasaka, *Isabella Bird*, 182.
69. Kennedy, *The Last Blank*, 208–12.
70. Schoenbauer Thurin, *Victorian Travelers*, 153.
71. Schoenbauer Thurin, *Victorian Travelers*, 153. See also Bays, *New History of Christianity*, 75–77; and Hsü, *Rise of Modern China*, 387–90.
72. Preston, *Boxer Rebellion*.
73. Bishop, *Yangtze Valley*, 210.
74. Isabella Bird to John Murray, July 6, 1896, MS 42028, NLS, fol. 24.
75. Chan, "Isabella Bird's Journey Valley," 224.
76. Bishop, *Yangtze Valley*, 210, 242.
77. Bishop, *Yangtze Valley*, 210.

78. Stoddart, *Life of Isabella Bird*, 314.
79. Bishop, *Yangtze Valley*, 261.
80. Bishop, *Yangtze Valley*, 209.
81. Bishop, *Yangtze Valley*, 209–10.
82. Bishop, *Yangtze Valley*, 231.
83. Schoenbauer Thurin, *Victorian Travelers*, 148.
84. Bishop, *Yangtze Valley*, 210.
85. Bishop, *Yangtze Valley*, 269.
86. Bird Bishop, *Korea and Her Neighbours*, 1:144, see also 98.
87. Bird Bishop, *Korea and Her Neighbours*, 1:144.
88. Schoenbauer Thurin, *Victorian Travelers*, 148–49.
89. Pratt, *Imperial Eyes*, 75–85.
90. Bishop, *Yangtze Valley*, 242.
91. Kennedy, *Last Blank Spaces*, 212.
92. Stanley, *Through the Dark Continent*, 1:231.
93. Stanley, *Through the Dark Continent*, 235–36; Jeal, *Stanley*, 178.
94. Driver, *Geography Militant*, 132.
95. "Henry M. Stanley," *Pall Mall Gazette*, February 11, 1878, 571; "Mr. Stanley's Explanations," *Saturday Review*, February 15, 1878, 207–8. On this incident, see Driver, *Geography Militant*, 132, 145.
96. "Mr. Stanley's Explanations," *Saturday Review*, February 15, 1878, 207–8.
97. Wisnicki, *Fieldwork of Empire*.
98. Driver, *Geography Militant*, 125.
99. For example, Barttelot, *Life*, 125, 154.
100. Larson, *Empire of Ice*.
101. Carpenter, *History of Vitamin C*, chapter 3.
102. Armston-Sheret, "Good Advertisement," 257–85.
103. Previous academic discussions of bodily examinations on polar expeditions have generally been by scholars interested in the history of medicine and extreme environment physiology. Heggie, "Why Isn't Exploration a Science?"; Heggie, *Higher and Colder*. See also Clements, *Science in an Extreme Environment*.
104. Scott, *Voyage*, 1:307–8.
105. Scott, *Voyage*, 1:308.
106. Armitage, *Two Years in the Antarctic*, 120.
107. Scott, *Voyage*, 1:308–9.
108. Scott, *Voyage*, 1:308.
109. Scott, *Voyage*, 1:308.
110. Scott, *Voyage*, 1:308; Armitage, *Two Years in the Antarctic*, 120–21.
111. Foucault, *Birth of the Clinic*.

112. Reginald Koettlitz to Robert Falcon Scott, October 3, 1902, MS 366/14;ER/23 SPRI Archives.

113. Jones, *Scott's Forgotten Surgeon*, 91.

114. Jones, *Scott's Forgotten Surgeon*, 104–5; Koettlitz, "Notes on the Galla of Walega and the Bertat," 50–55; Blundell and Koettlitz, "Journey through Abyssinia," 264–72, includes the same information. His presentation at the Anthropological Institute is noted in "3. Proceedings: Ordinary Meeting, February 13th, 1900," 3.

115. "Illustrated Interviews: Dr. Reginald Koettlitz," *South Polar Times*, June 1903, 46–48.

116. Armitage, *Two Years in the Antarctic*, 121; see also Mossakowski, "Sailors Dearly Love to Make Up," 200–202.

117. Jones, *Scott's Forgotten Surgeon*, 144–45.

118. Scott, *Voyage*, 1:309. See also Wilson, "Medical Aspect," 79; and Jones, *Scott's Forgotten Surgeon*, 145.

119. James William Lovibond, n.d. [1910], "Notebook Belonging to J. W. Lovibond on Eye and Hair Color for BAE 1910–13," RGS/RFS/5, RGS Archive.

120. Martin, "Tintometer."

121. Shackleton, *Heart of the Antarctic*, 1:162.

122. May and Lewis, "They Are Not the Ponies," 657. Scott's views in relation to this question are contested.

123. Heggie, *Higher and Colder*, 87.

124. Speke, *Journal of the Discovery*, 231.

125. Speke, *Journal of the Discovery*, 231.

126. Speke, "First Corrected Proofs of 'Journal of the Discovery of the Source of the Nile,'" NLS, fol. 145.

127. MacQueen, "Captain Speke's Discovery," 98.

7. SEXUAL BODIES

1. Speke to Secretary RGS, July 17, 1861, large text copy, JM5/53, RGS Archives, fol. 22. Bin Salim was initially employed as the headman on this expedition before being discharged at Tabora due to ill health. Speke, *Journal of the Discovery*, 614; see also Burton, *Lake Regions*, 1:9.

2. Speke, *Journal of the Discovery*, 37.

3. John Hanning Speke, 1863, "First Corrected Proofs of *Journal of the Discovery of the Source of the Nile*," MS 4873, NLS, fol. 233.

4. Driver, "Active Life"; Fabian, *Out of Our Minds*.

5. Hyam, *Empire and Sexuality*; Stoler, *Race, Education, and Desire*; Stoler, *Carnal Knowledge and Imperial Power*.

6. Arondekar, "Without a Trace"; Burton, *Burdens of History*; Howell, *Geographies of Regulation*; Legg, *Prostitution and the Ends of Empire*. On the regulation of sex in this period, see Foucault, *Will to Knowledge*.
7. Ghosh, *Sex and the Family*, 14–15.
8. Agnew, *Imperial Women Writers*, chapter 6.
9. Edward Said, *Orientalism*, 190.
10. McClintock, *Imperial Leather*, 22.
11. Bridges, "Nineteenth-Century East African Travel Records."
12. For example, Alrdich, *Colonialism and Homosexuality*, 29–34; Colligan, "Esoteric Pornography"; Colligan, "Race of Born Pederasts"; Kennedy, "Captain Burton's Oriental Muck Heap"; Phillips, "Writing Travel and Mapping Sexuality."
13. Burton, *Book of a Thousand Nights*, 10:206.
14. For differing perspectives see Kennedy, *Highly Civilized Man*, 30; and Lovell, *Rage to Live*, 88–89.
15. Kennedy, *Highly Civilized Man*, 239, 209, 169.
16. Brodie, *Devil Drives*, 105.
17. Burton, "Appendix IV"; for a history of this appendix and a discussion of its contents, see Brodie, *Devil Drives*, 110–11.
18. Burton, *Personal Narrative*, 1:87.
19. Said, *Orientalism*.
20. Burton, *Lake Regions*, 1:399.
21. Indeed, any private correspondence between the couple on this subject were destroyed; Lovell, *Rage to Live*, 381.
22. Kennedy, *Highly Civilized Man*, 213; Lovell, *Rage to Live*, xv.
23. Burton, *Life*; Burton, *Inner Life of Syria*; Burton, *Arabia, India, Egypt*.
24. Burton, *Life*, 1:74. This issue was brought to my attention by Kennedy, *Highly Civilized Man*, 90.
25. Kennedy, *Highly Civilized Man*, 90.
26. Jeal, *Livingstone*, 98.
27. Blaikie, *Personal Life of David Livingstone*, 297.
28. Jeal, *Livingstone*, 252–53.
29. David Livingstone quoted in Blaikie, *Personal Life of David Livingstone*, 299, 301.
30. Shipman, *To the Heart of the Nile*, 54. Her official date of birth was August 6, 1841, which would have made her nineteen at the time of the wedding.
31. Shipman, *To the Heart of the Nile*, 54.
32. Shipman, *To the Heart of the Nile*, 366.
33. Shipman, *To the Heart of the Nile*, 54.
34. Jeal, *Explorers of the Nile*.

35. "Mr. And Mrs. Baker's Nile Journey," *Eclectic Review*, August 1866, 108–24.
36. "Literature & Art," *Australasian*, October 13, 1866, 6.
37. "Literature & Art," 6; Florence was referred to as a "woman" despite being a teenager, as Samuel lied about her real age.
38. Murray and White, *Sir Samuel Baker*, 87.
39. Murray and White, *Sir Samuel Baker*, 50, 54, 87; Shipman, *To the Heart of the Nile*, 139.
40. Palsson, "Race and the Intimate," 373–74; see also Larson, *To the Edges of the Earth*.
41. Fabian, *Out of Our Minds*, 79–86.
42. Kennedy, *Last Blank Spaces*, 195–97, brought this issue to my attention.
43. Speke, "First Corrected Proofs," 224.
44. Speke, "First Corrected Proofs," 243, 231; Kennedy, *Last Blank Spaces*, 195–98.
45. Speke, "First Corrected Proofs," 242–43.
46. Kennedy, *Last Blank Spaces*, 198.
47. Speke, "First Corrected Proofs," 230–32.
48. Speke, "First Corrected Proofs," 234.
49. Jeal, *Explorers of The Nile*, 163.
50. Speke, "First Corrected Proofs," 233.
51. Speke, "First Corrected Proofs," 245.
52. Speke, "First Corrected Proofs," 234; see also 226.
53. Kennedy, *Last Blank Spaces*, 195–98; Jeal, *Explorers of the Nile*, 156–164; Finkelstein, "Unravelling Speke."
54. MacQueen, "Captain Speke's Discovery," 148
55. Kennedy, *Last Blank Spaces*, 196.
56. Keighren et al., *Travels into Print*.
57. Stanley "Journal S.A. 11, Full Transcript, (November 10, 1871—Unyanyembe, May 8, 1872)," in Leduc-Grimaldi and Newman, *Finding Dr. Livingstone*, 174.
58. William Bonny, "Autograph Journal of the Read Column," Royal Museum of Central Africa Archives (RMCA), HA.01.0098.69, entries March 1, 1888, March 3, 1888, and March 11, 1888.
59. William Bonny, Diary, Autograph Journal of the Read Column, Royal Museum of Central Africa Archives, HA.01.0098.69, entries March 1, 1888, and March 11, 1888.
60. Bonny Autograph Journal, HA.01.0098.69, entry March 1 and entry March 3, 1888.
61. Bonny, "Autograph Journal," HA.01.0098.69, RMCA, entry March 11, 1888.
62. Bonny, "Autograph Journal," HA.01.0098.70, RMCA, entry July 1, 1888, 5.
63. Bridges, "Nineteenth-Century East African Travel Records."

64. Stanley, *In Darkest Africa*, 1:470–519.
65. Kennedy, *Last Blank Spaces*, 172.
66. Stanley, *How I Found Livingstone*, 303.
67. Jeal, *Stanley*, 151.
68. Stanley, *Through the Dark*, 2:310.
69. Burton, *Supplementary Papers*, xxvii–viii.
70. Driver, *Geography Militant*, 146, 164.
71. Stanley, *My Kalulu*, 201.
72. Livingstone, "Romance of Slavery," 363.
73. Aldrich, *Colonialism and Homosexuality*, 39.
74. Livingstone, "Romance of Slavery," 363; see also Aldrich, *Colonialism and Homosexuality*, 46. For examples of this problematic approach, see McLynn, *Stanley*, 396; and Zambas, *Out of Africa*, 150. As Livingstone argues, some of these writers also overstate the sexual dimensions of Stanley's novel.
75. There is evidence that several nineteenth century travelers, most notably Alexander Von Humboldt, had "well-documented infatuations with and attractions to adult men"; Mackay et al., "Straight-Washing Ecological Legacies."
76. Waller, *Last Journals*, 2:200.
77. David Livingstone, "Field Diary XIV," Livingstone Online, entry May 28, 1872, https://livingstoneonline.org/in-his-own-words/catalogue?query=diary&view_pid=liv%3A000014&view_page=0.
78. "Livingstone and the Belle African," *Penny Illustrated Paper*, August 3, 1872, 75.
79. Henry M. Stanley, "Mr. Stanley's Paper Read Before the British Association at Brighton," *Ladies' Treasury*, October 1, 1872, 187; Stanley, *How I Found Livingstone*, 430.
80. Livingstone, "Field Diary XIV," entry May 28, 1872.
81. Livingstone, "Field Diary XIV," entry May 28, 1872. This is the very plausible motivation speculated by Gappah in her fiction retelling of the incident, *Out of Darkness*.
82. Livingstone, "Field Diary XIV," entry May 28, 1872.
83. Anson, *About Others and Myself*, 264. These four women stayed with the expedition to Cairo and then travelled to Zanzibar (via a circuitous route that left them stuck in Mauritius for a month).
84. Simpson, *Dark Companions*, 37.
85. Speke, *Journal of the Discovery*, 554; Simpson, *Dark Companions*, 37.
86. Grant, *Walk across Africa*, 257.
87. Speke, "First Corrected Proofs," 247, 249.
88. Speke, "First Corrected Proofs," 260; her age is mentioned on page 224.

89. Grant, *Walk across Africa*, 257.
90. Speke, "First Corrected Proofs," 248–49.
91. Rockel, *Carriers of Culture*, 121, 127.
92. Speke, *Journal of the Discovery*, 190.
93. Rockel, *Carriers of Culture*, 128.
94. Thomson, *To the Central African Lakes*, 1:263–64.
95. Birkett, *Spinsters Abroad*.
96. Marianne North Letter to Arthur Burnell, January 20, 1878, quoted in Agnew, "Old Vagabond."
97. Chubbuck, "Introduction," in *Letters to Henrietta*, 15–17.
98. Isabella L. Bird to John Murray, 1881, January 9, MS.42025, NLS, 67.
99. Blunt, *Travel Gender and Imperialism*, 68–69.
100. Birkett, *Spinsters Abroad*, 156.
101. Bird, *Lady's Life*, 91.
102. Isabella Bird to Henrietta Bird in Bird, November 18–20, 1872, in *Letters to Henrietta*, 175.
103. Kingsley, *Notes on Sport and Travel*, 174–6.
104. Gurney and Podmore, *Phantasms of the Living*, 531–32.
105. "Current Literature," *Argus*, February 28, 1880, 4.
106. "A Lady's Life in the Rocky Mountains," *John Bull*, November 29, 1879, 71.
107. The book was noted in as one of the "women's books of the month" in the feminist journal the *Englishwoman's Review*, February 14, 1880, 85. But it was also a source of advice for future settlers. "Multiple News Items," *Standard*, December 26, 1879, 8.
108. Wheeler, *Terra Incognita*, 51.
109. Betty Robson to Ernest Shackleton, July 2, 1921, MS1591/1/11/16, Shackleton-Rowett Antarctic Expedition Collection, SPRI; on his 1914–17 expedition, see Spufford, *I May Be Some Time*, 144.
110. Fiennes, *Shackleton*, 201, 131.
111. Jones, "Truth about Captain Scott," 859.
112. Seligman, *Rum, Sodomy, Prayers*, 65–67. The navy made growing efforts to punish homosexual activities in the run up to World War I.
113. Seligman, *Rum, Sodomy, Prayers*, 68, 89. Officers were particularly unlikely to face trial, as the Navy was keen to protect the reputation of the service.
114. Shackleton, *Heart of the Antarctic*, 1:143.
115. Fiennes, *Shackleton*, 73.
116. Jackson, "Three Years' Exploration," 129.
117. Frederick Jackson Diary, February 8–September 2, 1897, MS/287/1/3; BJ, SPRI, entry June 30, 1897. The passage was crossed out and did not appear in his published book.

118. Armston-Sheret, "Leaky Bodies."
119. Scott, *Voyage*, 1:379.
120. Edward Wilson, "Journal, June 1, 1910, to February 27, 1912," typed bound copy, MS715/2; BJ, 40, July 15, 1910.
121. Fiennes, *Shackleton*, 183.
122. Strange, "Reconsidering the 'Tragic,'" 74.
123. Mossakowski, "Sailors Dearly Love to Make Up" 230.
124. Scott, *Voyage*, 1:379–80.
125. Mossakowski, "Sailors Dearly Love to Make Up," 285.
126. Mossakowski, "Sailors Dearly Love to Make Up," 249.
127. Scott, *Voyage*, 1:381; Mossakowski, "Sailors Dearly Love to Make Up," 286.

CONCLUSION

1. David Livingstone, "Field Diary XVII, April 9–April 27, 1873," entry April 10, 1873, fol. 15, Livingstone Online, https://livingstoneonline.org/in-his-own-words/catalogue?query=field+diary&view_pid=liv%3A000017&view_page=0.
2. Livingstone, *Last Journals*, 2:316.
3. Scott, *Journals*, 459.
4. Pettitt, *Dr. Livingstone, I Presume?*, 129. This said, some elements of the public remained fascinated by graphic descriptions of Livingstone's "mangled corpse"; Livingstone, *Livingstone's Lives*, 103.
5. Kristeva, *Powers of Horror*, 3.
6. Longhurst, *Bodies*.
7. MacKenzie, "David Livingstone," 216.
8. Barczewski, *Antarctic Destinies*.
9. Margot and Capparell, *Shackleton's Way*.
10. Simmons, *Terror*.
11. Thiess, "Dan Simmons's *The Terror*."
12. Moore, "Comparing Colonialisms." Simmon's book is generally less sensitive in its portrayal of Inuit people than the AMC series.
13. For example, Livingstone, *Livingstone's Lives*; Rupke, *Alexander Von Humboldt*.
14. Said, *Culture and Imperialism*, xiii.
15. Graham Burnett, "It Is Impossible"; Whatmore, *Hybrid Geographies*.
16. Sivasundaram, "Imperial Transgressions."
17. Stoler, *Along the Archival Grain*, 47.
18. Bridges, "Explorers' Texts," 72, 79; Bridges, "Nineteenth-Century East African Travel Records," 180–86.
19. Kennedy, *Last Blank Spaces*, 249.
20. Scott, *Voyage*, 1:499–500.

21. See Robert F. Scott, "Draft Fragment of Chapter 18 of the Voyage of the Discovery (London, 1905)," MS 1453/24;D, SPRI, fols. 5, 7.
22. Kingsley, *Travels in West Africa*, 100, 257.
23. Wisnicki, *Fieldwork of Empire*, 1–18.
24. Stoler, *Duress*; Nicolaïdis et al., *Echoes of Empire*.
25. Jazeel et al. "Decolonising Spaces," 517; Johnson, "Academic Witness," 23.
26. Driver, "Active Life"; Hevly, "Heroic Science."
27. Anadu et al., "Ten Steps."
28. Quality Assurance Agency, "Subject Benchmark Statement: Geography," 10. For differing perspectives on this subject, see Guasco, "On the Ethics of Not Going There"; and Howkins, "Have You Been There?"
29. Maguire, "Gender Differences," 213.
30. That said, some women find enjoyment in fieldwork. See Bracken (née Bull) and Mawdsley, "'Muddy Glee'; and" Pang, "Gender, Culture, and Astrophysical," 34.
31. Leyland et al., "Classics Revisited"; Tucker and Horton, "The Show Must Go On!"
32. Lawrence and Dowley "Six Simple Steps," 53; see also the RGS-IBG, "Principles for Undergraduate Fieldwork."
33. North et al., "Out of Africa"; Rogers et al., "Geology Uprooted!"
34. Hirsch, "Mental Health, Coloniality," 211.
35. Tang and Yuck, "Decolonisation Is Not a Metaphor."
36. Esson, "Why and the White."
37. Desai, "Black and Minority Ethnic (BME) Student and Staff."
38. Shapin, "Invisible Technician," 562–63.

BIBLIOGRAPHY

ARCHIVES AND COLLECTIONS

American Geographical Society Library, University of Wisconsin, Milwaukee.

American Geographical Society of New York Records, 1723–2010, bulk 1854–2000.

Charles P. Daley correspondence blocks, Q–R.

Huntington Library.

Richard Francis Burton Library Manuscripts Collection.

National Library of Scotland (NLS).

Blackwood Papers, "Correspondence with Speke and Manuscripts and Proofs of Journal of the Discovery of the Source of the Nile," Mss. 4143–4873.

John Murray Collection, particularly Isabella L. Bishop's correspondence with the firm: Mss. 42020–42029.

Miscellany of Letters and Documents of African Explorers, Missionaries and Administrators, Ms. 20311.

Papers of and Concerning Isabella Bishop, 1855–1904, Mss. 42020–42036.

Papers of James Augustus Grant and His Family, Mss. 17901–17926.

National Maritime Museum Archives.

Albert Markham Collection: MRK.

Heraldry Collection.

Royal Asiatic Society Archives.

Richard Francis Burton Collection.

Royal Geographical Society (with the Institute of British Geographers) Archives (RGS).

Admission of Women Fellows: RGS/AP/93/1.

Correspondence Blocks 1851–60; 1861–70; 1888–98; Isabella Lucy Bishop; James Augustus Grant; John Hanning Speke; Richard Francis Burton: RGS/CB.

Speke, John Hanning. 1857. "12 Sketch Maps Showing the Route of a Journey by Speke and Burton, to and around Sea Ujiji." Map Room Collection, CU8-AFE-TZE-S507.

John Hanning Speke Collection: RGS/JHS.
Reginald Koettlitz Collection: RGS/LMS K 12.
National Antarctic Expedition Collection: RGS/AA.
Richard Francis Burton Collection: RGS/RFB.
Robert Falcon Scott Collection: RGS/RFS.
Unpublished Journal Manuscripts Collection.

Royal Museum for Central Africa, Tervuren, Belgium.
Henry Morton Stanley Archives.

Royal Society.
Manuscripts General.

Scott Polar Research Institute (SPRI), University of Cambridge.
British Antarctic Expedition Collection.
British National Antarctic Expedition Collection.
Edward Adrian Wilson Collection.
Frederick Jackson Collection.
Reginald Koettlitz Collection.
Shackleton-Rowett Antarctic Expedition Collection.
South Polar Times Collection, GB 15 South Polar Times.
William Lashly Collection.

Adam Matthew digital archive, particularly:
Age of Exploration Collection.
Joseph John Kinsey Collections, Alexander Turnbull Library: National Library of New Zealand.
Nineteenth-Century Literary Society Collection.

Auckland Public Library.
Letters to Sir George Grey, Sir George Grey Collection.

Livingstone Online.

PUBLISHED SOURCES

"100 Great British Heroes," *BBC News*, August 21, 2002. http://news.bbc.co.uk/1/hi/entertainment/tv_and_radio/2208671.stm.

"3. Proceedings: Ordinary Meeting, February 13th, 1900." *Journal of the Anthropological Institute of Great Britain and Ireland*, no. 30 (1900): 3.

Adair, Gigi. "The 'Feringhi Hakīm': Medical Encounters and Colonial Ambivalence in Isabella Bird's Travels in Japan and Persia." *Studies in Travel Writing* 21, no. 1 (January 2, 2017): 63–75.

Agnew, Eadaoin. "'An Old Vagabond': Science and Sexuality in Marianne North's Representations of India." *Nineteenth-Century Gender Studies* 7, no. 2 (Summer 2011). https://www.ncgsjournal.com/issue72/agnew.html.

Aldrich, Robert. *Colonialism and Homosexuality*. London: Routledge, 2003.

Alp, Bill. "Commentary on 'Could Captain Scott Have Been Saved?' Cecil Meares and the 'Second Journey' That Failed." *Polar Record* 54, no. 2 (2018): 181–83.

———. "Dogs of the British Antarctic Expedition, 1910–13." *Polar Record* 55, no. 6 (2019): 476–96.

Amundsen, Roald. *The South Pole: An Account of the Norwegian Antarctic Expedition in the "Fram," 1910–12*. Translated by A. G. Chater. 2 vols. London: John Murray, 1912.

Anadu, Joshua, Hendratta Ali, and Christopher Jackson. "Ten Steps to Protect BIPOC Scholars in the Field." Eos, November 10, 2020. https://eos.org/opinions/ten-steps-to-protect-bipoc-scholars-in-the-field.

Anderson, Kay and Colin Perrin. "How Race Became Everything: Australia and Polygenism." *Ethnic and Racial Studies* 31, no. 5 (2008): 962–90.

Anson, Archibald Edward Harboed. *About Others and Myself, 1745 to 1920*. London: John Murray, 1920.

Anthony, Jason C. *Hoosh: Roast Penguin, Scurvy Day, and Other Tales of Antarctic Cuisine*. Lincoln: University of Nebraska Press, 2013.

Ardeleanu, Constantin. "A British Meat Cannery in Moldavia (1844–52)." *Slavonic and East European Review* 90, no. 4 (2012): 671–704.

Armitage, Albert. *Two Years in the Antarctic: Being a Narrative of the British National Antarctic Expedition*. London: Edward Arnold, 1905.

Armston-Sheret, Edward. "Diversifying the Historical Geography of Exploration: Subaltern Body Work on British-Led Expeditions c. 1850–1914." *Journal of Historical Geography* 80 (April 2023): 58–68.

———. "'A Good Advertisement for Teetotalers': Polar Explorers and Debates over the Health Effects of Alcohol, 1875–1904." *Social History of Alcohol and Drugs* 33, no. 2 (2019): 257–85.

———. "Leaky Bodies: Syphilitic Incontinence on the Jackson-Harmsworth Expedition." *Victorian Review* 45, no. 1 (2019): 10–14.

———. "Recentering the Body in Histories of British Exploration." PhD diss., Royal Holloway, University of London, 2021.

———. "Tainted Bodies: Scurvy, Bad Food, and the Reputation of the British National Antarctic Expedition, 1901–4." *Journal of Historical Geography* 65 (2019): 23–28.

Arnold, David. *Colonizing the Body: State Medicine and Epidemic Disease in Nineteenth-Century India*. Berkeley: University of California Press, 1993.

———. "Envisioning the Tropics." In *Tropical Visions in an Age of Empire*, edited by Felix Driver and Luciana Martins, 137–55. Chicago: University of Chicago Press, 2005.

———. *The Tropics and the Traveling Gaze: India, Landscape, and Science, 1800–1856*. Seattle: University of Washington Press, 2006.

Arondekar, Anjali. "Without a Trace: Sexuality and the Colonial Archive." *Journal of the History of Sexuality* 14, nos. 1–2 (2005): 10–27.

Asad, Talal, ed. *Anthropology and the Colonial Encounter*. London: Ithaca, 1975.

Augustine. *Confessions*. Translated by Henry Chadwick. Oxford: Oxford University Press, 1992.

Baigent, Elizabeth. "'Deeds Not Words?' Life Writing and Early Twentieth-Century British Polar Exploration." In *New Spaces of Exploration: Geographies of Discovery in the Twentieth Century*, edited by Simon Naylor and James Ryan, 23–51. London: I. B. Tauris, 2010.

Baker, Samuel W. *Albert Nyanza: Great Basin of the Nile*. 2 vols. London: Sidgwick, 1866.

———. *The Rifle and the Hound in Ceylon*. London: Longman, Brown, Green & Longmans, 1854.

Baldwin, Peter C. "How Night Air Became Good Air, 1776–1930." *Environmental History* 8, no. 3 (2003): 412–29.

Ballantyne Tony, and Antoinette Burton, eds. *Bodies in Contact: Rethinking Colonial Encounters in World History*. Durham NC: Duke University Press, 2005.

Barczewski, Stephanie. *Antarctic Destinies: Scott, Shackleton, and the Changing Face of Antarctic Heroism*. London: Continuum, 2007.

———. *Heroic Failure and the British*. New Haven CT: Yale University Press, 2016.

Barnes, David S. "The Senses in Medicine: Seeing, Hearing, and Smelling Disease." In *A Cultural History of the Senses in the Age of Empire*, edited by Constance Classen, vol. 5, 137–60. London: Bloomsbury, 2014.

Barnett, Clive. "Impure and Worldly Geography: The Africanist Discourse of the Royal Geographical Society, 1831–73." *Transactions of the Institute of British Geographers* 23, no. 2 (1998): 239–51.

Barr, Pat. *A Curious Life for a Lady: The Story of Isabella Bird, Traveller Extraordinary*. London: Penguin, 1985.

Barrett-Gaines, Kathryn. "Travel Writing, Experiences, and Silences: What Is Left Out of European Travelers' Accounts: The Case of Richard D. Mohun." *History in Africa* 24 (1997): 53–70.

Bartky, Sandra Lee. "Foucault, Femininity and the Modernization of Patriarchal Power." In *The Politics of Women's Bodies: Sexuality, Appearance, and Behaviour*, edited by Rose Weitz, 25–45. Oxford: Oxford University Press, 1998.

Barttelot, Edmund Musgrave, and Walter George Barttelot. *The Life of Edmund Musgrave Barttelot, Captain and Brevet-Major Royal Fusiliers, Commander of the Rear Column of the Emin Pasha Relief Expedition; Being an Account of His Services for the Relief of Kandahar, of Gordon, and of Emin, from His Letters and Diary*. London: R. Bentley & Son, 1890.

Bates, Henry Walter. "Natural History." In *Hints to Travellers: Scientific and*

General, 5th ed, edited by Henry Godwin Austen, John Knox Laughton, and Douglas Freshfield, 203–21. London: Royal Geographical Society, 1883.
———. *A Naturalist on the River Amazons*. 2 vols. London: John Murray 1863.
Baughman, T. H. *Pilgrims on The Ice: Robert Falcon Scott's First Antarctic Expedition*. Lincoln: University of Nebraska Press, 1999.
Bays, Daniel H. *A New History of Christianity in China*. Oxford: Wiley-Blackwell, 2012.
Beattie, James. *Empire and Environmental Anxiety: Health, Science, Art, and Conservation in South Asia and Australasia, 1800–1920*. Basingstoke: Palgrave Macmillan, 2011.
Beesly, Augustus Henry. *Sir John Franklin*. London: Marcus Ward, 1881.
Begiato, Joanne. *Manliness in Britain, 1760–1900: Bodies, Emotion, and Material Culture*. Manchester: Manchester University Press, 2020.
Bell, Leonard. "Eyeing Samoa: People, Places, and Spaces in Photographs of the Late Nineteenth and Early Twentieth Centuries." In *Tropical Visions in an Age of Empire*, edited by Felix Driver and Luciana Martins, 156–74. Chicago: University of Chicago Press, 2005.
Bell, Morag, and Cheryl McEwan. "The Admission of Women Fellows to the Royal Geographical Society, 1892–1914: The Controversy and the Outcome." *Geographical Journal* 162, no. 3 (1996): 295–312.
Berenson, Edward. *Heroes of Empire: Five Charismatic Men and the Conquest of Africa*. Berkeley: University of California Press, 2011.
Berger, John. *Ways of Seeing*. London: Penguin, 1971.
Bird, Isabella L. *The Golden Chersonese and the Way Thither*. London: John Murray, 1883.
———. *The Hawaiian Archipelago: Six Months among the Palm Groves, Coral Reefs, and Volcanoes of the Sandwich Islands*. London: John Murray, 1875.
———. *A Lady's Life in the Rocky Mountains*. London: John Murray, 1879.
———. *A Lady's Life in the Rocky Mountains*. 4th ed. London: John Murray, 1881.
———. *Letters to Henrietta*. Edited by Kay Chubbuck. Boston: Northeastern University Press, 2002.
———. *Unbeaten Tracks in Japan: An Account of Travels on Horseback in the Interior, Including Visits to the Aborigines of Yezo and the Shrine of Nikó and Isé*. 2 vols. London: John Murray, 1880.
———. *Unbeaten Tracks in Japan: An Account of Travels in the Interior, Including Visits to Aborigines of Yezo and the Shrine of Nikkô*. London: John Murray, 1905.
Bird Bishop, Isabella. *Among the Tibetans*. London: Religious Tract Society, 1894.
———. *Korea and Her Neighbours: A Narrative of Travel with an Account of the Recent Vicissitudes and Present Position of the Country*. 2 vols. London: John Murray, 1898.

Birkett, Dea. *Spinsters Abroad: Victorian Lady Explorers*. London: Sutton, 2004.

Bishop, Mrs. J. F. [Isabella Bird]. "Anniversary Address: The Upper Karun Region and the Bakhtiari Lurs." *Scottish Geographical Magazine* 8, no.1 (1892): 1–14.

———. *Chinese Pictures: Notes on Photographs Made in China*. London: Cassel, 1900.

———. *Journeys in Persia and Kurdistan: Including a Summer in the Upper Karun Region and a Visit to the Nestorian Rayahs*. 2 vols. London: John Murray, 1891.

———. "A Journey through Lesser Tibet." *Scottish Geographical Magazine* 8, no. 10 (1892): 513–28.

———. *The Yangtze Valley and Beyond: An Account of Journeys in China, Chiefly in the Province of Sze Chaun and among the Man-Tze of the Somo Territory*. London: John Murray, 1899.

Blaikie, William Garden. *The Personal Life of David Livingstone: Chiefly from His Unpublished Journals and Correspondence in the Possession of His Family*. London: John Murray, 1880.

Bloom, Lisa. *Gender on Ice: American Ideologies of Polar Expedition*. Minneapolis: University of Minnesota Press, 1993.

Blundell, Herbert Weld, and Reginald Koettlitz. "A Journey through Abyssinia to the Nile (Continued): Notes on Geology and Anthropology." *Geographical Journal* 15, no. 3 (1900): 264–72.

Blunt, Alison. *Travel, Gender, and Imperialism: Mary Kingsley and West Africa*. New York: Guilford. 1994.

Boisseron, Bénédicte. *Afro-Dog: Blackness and the Animal Question*. New York: Columbia University Press, 2018.

Bourke, Joanna. *The Story of Pain: From Prayers to Painkillers*. Oxford: Oxford University Press, 2014.

Bracken (née Bull), Louise, and Emma Mawdsley. "'Muddy Glee': Rounding out the Picture of Women and Physical Geography Fieldwork." *Area* 36, no. 3 (2004): 280–86.

Brescius, Moritz von. *German Science in the Age of Empire*. Cambridge: Cambridge University Press, 2018.

Bressey, Caroline. "Surfacing Black and Brown Bodies in the Digital Archive: Domestic Workers in Late Nineteenth-Century Australia." *Journal of Historical Geography* 70 (2020): 1–11.

Bridges, Roy C. "Explorers' Texts and the Problem of Reactions by Non-Literate Peoples." *Studies in Travel Writing* 2, no. 1 (1998): 65–84.

———. "Nineteenth-Century East African Travel Records with an Appendix on 'Armchair Geographers' and Cartography." *Paideuma* 33 (1987): 179–96.

Brierley, Jean, and Thomas Spear. "Mutesa, The Missionaries, and Christian

Conversion in Buganda." *International Journal of African Historical Studies* 21, no. 4 (1988): 601–18.
Brodie, Fawn. *The Devil Drives: A Life of Sir Richard Burton*. London: Eyre & Spottiswode, 1967.
Buller, Henry. "Animal Geographies I." *Progress in Human Geography* 38, no. 2 (2012), 308–18.
Burnett, Graham, D. *Masters of All They Surveyed: Exploration, Geography, and a British El Dorado*. Chicago: University of Chicago Press, 2000.
Burton, Antoinette M. "The Body In/As World History." In *A Companion to World History*, edited by Douglas Northop, 272–384. Oxford: Wiley-Blackwell, 2012.
———. *Burdens of History: British Feminists, Indian Women, and Imperial Culture, 1865–1915*. Chapel Hill: University of North Carolina Press, 1994.
Burton, Antoinette M., and Renisa Mawani, eds. *Animalia: An Anti-Imperial Bestiary for Our Times*. Durham NC: Duke University Press, 2020.
Burton, Isabel. *Arabia, Egypt, India: A Narrative of Travel*. London: W. Mullan & Son, 1879.
———. *The Inner Life of Syria, Palestine, and the Holy Land: From My Private Journal*. London: Henry King, 1875.
———. *The Life of Captain Sir Richard F. Burton*. 2 vols. London: Chapman & Hall, 1893.
Burton, Richard F. "Appendix IV: A Brief Description of Certain Peculiar Customs." Privately published, 1856. http://burtoniana.org/books/1856-First%20footsteps%20in%20east%20africa/appendixiv/index.html.
———. *First Footsteps in East Africa: An Exploration of Harar*. London: Longman Brown, Green & Longmans, 1856.
———. *The Lake Regions of Central Africa: A Picture of Exploration*. 2 vols. London: Longman, Green, Longman & Roberts, 1860.
———. "The Lake Regions of Central Equatorial Africa, with Notices of the Lunar Mountains and the Sources of the White Nile; Being the Results of an Expedition Undertaken under the Patronage of Her Majesty's Government and the Royal Geographical Society of London, in the Years 1857–1859." *Journal of the Royal Geographical Society* 29 (1859): 1–454.
———. *Personal Narrative of a Pilgrimage to El-Medinah and Meccah*. 3 vols. London: Longman, Brown & Green, 1855.
———. *Supplementary Papers to the Mwata Cazembe*. Privately published, 1873.
———. *Two Trips to Gorilla Land and the Cataracts of the Congo*. 2 vols. London: Sampson, Low, Marston & Searle, 1876.
———. *Zanzibar: City, Island, and Coast*. 2 vols. London: Tinsley Brothers, 1872.
Burton, Richard F., trans. *The Book of a Thousand Nights and a Night*. 10 vols. London: privately published, 1886.

Burton, Richard F., and James MacQueen. *The Nile Basin.* London: Tinsley Brothers, 1864.

Butler, Judith. *Bodies That Matter: On the Discursive Limits of "Sex."* New York: Routledge, 1993.

———. *Gender Trouble: Feminism and the Subversion of Identity.* Routledge Classics ed. New York: Routledge, [1990] 2007.

Bynum, Caroline. "Why All the Fuss about the Body? A Medievalist's Perspective." *Critical Inquiry* 22, no. 1 (1995): 1–33.

Caldwell, Sabrina Bleecker. "Framing Digital Image Credibility: Image Manipulation Problems, Perceptions and Solutions." PhD diss., Australian National University, 2016.

Cameron, Verney Lovett. *Across Africa.* London: Daldy, Ibister, 1877.

Campbell, Joseph. *The Hero with a Thousand Faces.* 3rd ed. Novato CA: New World Library, [1949] 2008.

Campbell Davidson, Lilias. *Hints to Lady Travellers: At Home and Abroad.* London: Iliffe & Son, 1889.

Canning, Kathleen. "The Body as Method? Reflections on the Place of the Body in Gender History." *Gender & History* 11, no. 3 (1999): 499–513.

Carlyle, Thomas. *On Heroes, Hero-Worship, and the Heroic in History.* London: U. P. James, 1842.

Carnochan, W. B. *The Sad Story of Burton, Speke, and the Nile; or Was John Hanning Speke a Cad?* Stanford CA: Stanford University Press, 2006.

Carpenter, Kenneth. *A History of Vitamin C and Scurvy.* Cambridge: Cambridge University Press, 1986.

Carter, Paul. *The Road to Botany Bay: An Exploration of Landscape and History.* London: Faber & Faber, 1987.

Casada, James A. "James A. Grant and the Royal Geographical Society." *Geographical Journal* 140, no. 2 (1974): 245–53.

Cathcart, Michael. *Starvation in a Land of Plenty: Wills' Diary of the Fateful Burke and Wills Expedition.* Canberra: National Library Australia, 2013.

Cavell, Janice. "Publishing Sir John Franklin's Fate: Cannibalism, Journalism, and the 1881 Edition of Leopold McClintock's 'The Voyage of the *Fox* in the Arctic Seas.'" *Book History* 16 (2013): 155–84.

———. *Tracing the Connected Narrative: Arctic Exploration in British Print Culture, 1818–1860.* Toronto: University of Toronto Press, 2008.

Chakrabarty, Dipesh. "A Small History of Subaltern Studies." In *A Companion to Postcolonial Studies*, edited by Henry Schwarz and Sangeeta Ray, 467–85. Malden, MA: Blackwell, 2000.

Chan, May Caroline. "Isabella Bird's Journey through the Yangtze Valley: Victorian Travel Narrative as a Historical Record of British Imperial Desires in

China." In *Clio's Daughters: British Women Making History, 1790–1899*, edited by Lynette Felber, 215–35. Newark: University of Delaware Press, 2007.

Chang, Elizabeth Hope. *Britain's Chinese Eye: Literature, Empire, and Aesthetics in Nineteenth-Century Britain.* Stanford CA: Stanford University Press, 2010.

Chantiluke, Roseanne, Brian Kwaboa, and Athinangamso Nkopo, eds. *Rhodes Must Fall: The Struggle to Decolonise the Racist Heart of Empire.* London: Zed, 2018.

Cherry-Garrard, Apsley. *The Worst Journey in the World.* 2 vols. London: Constable, 1922.

———. *The Worst Journey in the World.* London: Picador, 1994.

Clarence-Smith, William G. "Equids across the Indian Ocean c. 1800–1918." *Journal for Maritime Research* 22, nos. 1–2 (2020): 41–57.

Clark, James. *The Influence of Climate in the Prevention and Cure of Chronic Diseases.* London: John Murray, 1829.

Clayton, Daniel. *Islands of Truth: The Imperial Fashioning of Vancouver Island.* Vancouver: University of British Columbia Press, 2000.

Clements, Philip W. *Science in an Extreme Environment: The 1963 American Mount Everest Expedition.* Pittsburgh PA: University of Pittsburgh Press, 2018.

Coen, Deborah, R. *Climate in Motion: Science, Empire, and the Problem of Scale.* Chicago: University of Chicago Press, 2018.

Colligan, Colette. "'Esoteric Pornography': Sir Richard Burton's Arabian Nights and the Origins of Pornography." *Victorian Review* 28, no. 2 (2002): 31–64.

———. "'A Race of Born Pederasts': Sir Richard Burton, Homosexuality, and the Arabs." *Nineteenth-Century Contexts* 25, no. 1 (March 1, 2003): 1–20.

Cooter, Roger. "The Turn of the Body: The History and Politics of the Corporeal." *Arbor* 186, no. 743 (2010): 393–405.

Cosgrove, Denis. *Geography and Vision: Seeing, Imagining, and Representing the World.* London: I. B. Tauris, 2008.

Cowan, Thomas. *Subaltern Frontiers: Agrarian City-Making in Gurgaon.* Cambridge: Cambridge University Press, 2022.

Cox, Natalie. "Armchair Geography: Speculation, Synthesis, and the Culture of British Exploration, c. 1830–c. 1870." PhD diss., University of Warwick, 2016.

Craciun, Adriana. "What Is an Explorer?" *Eighteenth-Century Studies* 45, no. 1 (2011): 29–51.

———. *Writing Arctic Disaster: Authorship and Exploration.* Cambridge: Cambridge University Press, 2016.

Craggs, Ruth. "Decolonising the Geographical Tradition." *Transactions of the Institute of British Geographers* 44, no. 3 (2019): 444–46.

Crang, Mike. "Qualitative Methods: Touchy, Feely, Look-See?" *Progress in Human Geography* 27, no. 4 (2003): 494–504.

Creighton, Louise. *Some Famous Women.* London: Longmans, Green, 1909.

Cribb, Robert, Helen Gilbert, and Helen Tiffin. *Wild Man from Borneo: A Cultural History of the Orangutan*. Honolulu: University of Hawai'i Press, 2014.

Cubitt, Geoffrey, and Alan Warren, eds. *Heroic Reputations and Exemplary Lives*. Manchester: Manchester University Press, 2000.

Darwin, Charles, ed., *The Zoology of the Voyage of H.M.S. Beagle, under the Command of Captain Fitzroy, R.N., during the Years 1832 to 1836*. London: Smith & Elder, 1839.

Davenport, Adam W. H. *Celebrated Women Travelers of the Nineteenth Century*. New York: Dutton, 1903.

David, Alison Matthews. "Elegant Amazons: Victorian Riding Habits and the Fashionable Horsewoman." *Victorian Literature and Culture* 30, no. 1 (2002): 179–210.

Davis, Wade. *Into the Silence: The Great War, Mallory, and the Conquest of Everest*. London: Bodley Head, 2011.

Della Dora, Veronica, and Cosgrove, Dennis E., eds. *High Places: Cultural Geographies of Mountains and Ice*. London: I. B. Tauris, 2008.

Derrida, Jacques. *The Animal That Therefore I Am*. Translated by David Wills. New York: Fordham University Press, 2008.

———. *Archive Fever: A Freudian Impression*. Translated by Eric Prenowitz. Chicago: University of Chicago Press, 1998.

Desai, Vandana. "Black and Minority Ethnic (BME) Student and Staff in Contemporary British Geography." *Area* 49, no. 3 (2017): 320–23.

Dettelbach, Michael. "Global Physics and Aesthetics Empire: Humboldt's Physical Portrait of the Tropics." In *Visions of Empire: Voyages, Botany, and Representations of Nature*, edited by David P. Miller and Peter H. Reill, 258–92. Cambridge: Cambridge University Press, 1996.

Dick, Lyle. 2001 *Muskox Land: Ellesmere Island in the Age of Contact*. Calgary: University of Calgary Press, 2001.

Dodds, Klaus. *Pink Ice: Britain and the South Atlantic Empire*. London: I. B. Tauris, 2002.

Domosh, Mona. "Beyond the Frontiers of Geographical Knowledge." *Transactions of the Institute of British Geographers* 16, no. 4 (1991): 488–90.

———. "Toward a Feminist Historiography of Geography." *Transactions of the Institute of British Geographers* 16, no. 1 (1991): 95–104.

Dritsas, Lawrence. "Expeditionary Science: Conflicts of Method in Mid-Nineteenth-Century Geographical Discovery." In *Geographies of Nineteenth-Century Science*, edited by Charles W. J. Withers and David Livingstone, 255–78. Chicago: University of Chicago Press, 2011.

———. "From Lake Nyassa to Philadelphia: A Geography of the Zambesi Expedition, 1858–64." *British Journal for the History of Science* 38, no. 1 (March 2005): 35–52.

———. *Zambesi: David Livingstone and Expeditionary Science in Africa*. London: I. B. Tauris 2010.

Driver, Felix. "The Active Life: The Explorer as Biographical Subject." In *Oxford Dictionary of National Biography*. Online edition, 2005. http://www.oxforddnb.com (accessed May 2, 2024).

———. "Distance and Disturbance: Travel, Exploration, and Knowledge in the Nineteenth Century." *Transactions of the Royal Historical Society* 14, no. 1 (2005): 73–92.

———. "Exploration as Knowledge Transfer: Exhibiting Hidden Histories." In *Mobilities of Knowledge*, edited by Heike Jöns, Peter Meusburger, and Michael Heffernan, 85–104. Cham: Springer International, 2017.

———. "Face to Face to with Nain Singh: The Schlagintweit Collections and Their Uses." In *Naturalists in the Field: Collecting, Recording, and Preserving the Natural World from the Fifteenth to the Twenty-First Century*, edited by Arthur MacGregor, 441–69. Leiden: Brill, 2018.

———. *Geography Militant: Cultures of Exploration and Empire*. Oxford: Wiley-Blackwell, 2001.

———. "Geography's Empire: Histories of Geographical Knowledge." *Environment and Planning D: Society and Space* 10, no. 1 (1992): 23–40.

———. "Henry Morton Stanley and His Critics: Geography, Exploration, and Empire." *Past and Present* 133, no.1 (1991): 134–67.

———. "Hidden Histories Made Visible? Reflections on a Geographical Exhibition." *Transactions of the Institute of British Geographers* 38, no. 3 (2013): 420–35.

———. "Material Memories of Travel: The Albums of a Victorian Naval Surgeon." *Journal of Historical Geography* 69 (2020): 32–54.

———. "Power, Space, and the Body: A Critical Assessment of Foucault's *Discipline and Punish*." *Environment and Planning D: Society and Space* 3, no. 4 (December 1, 1985): 425–46.

Driver, Felix, and Lowri Jones. *Hidden Histories of Exploration*. London: Royal Geographical Society, 2009.

Driver, Felix, and Luciana Martins. "John Septimus Roe and the Art of Navigation, c. 1815–1830." *History Workshop Journal* 54, no. 1 (October 1, 2002): 144–61.

Driver, Felix, and Luciana Martins, eds. *Tropical Visions in an Age of Empire*. Chicago: University of Chicago Press, 2005.

Driver, Felix, Mark Nesbitt, and Caroline Cornish, eds. *Mobile Museums: Collections in Circulation*. London: UCL, 2021.

Eastlake, Elisabeth. "Lady Travellers." *Quarterly Review* 76 (1845): 98–99.

Elias, Norbert. *The Civilizing Process: Sociogenetic and Psychogenetic Investigations*. Rev. ed. Translated by Edmund Jephcott. Oxford: Blackwell, 2000.

Elliot, Andrew. "Ito and Isabella in the Contact Zone: Interpretation, Mimicry, and *Unbeaten Tracks in Japan*." *Electronic Journal of Contemporary Japanese Studies* (December 2008). http://www.japanesestudies.org.uk/articles/2008/Elliott.html.

———. "'It's Japan, but Yet There is a Difference Somehow': Editorial Change and Yezo in Isabella Bird's *Unbeaten Tracks in Japan*." *Journeys* 9, no. 1 (2008): 1–20.

Esson, James. "'The Why and the White': Racism and Curriculum Reform in British Geography." *Area* 52, no. 4 (2020): 708–15.

Esson, James, Patricia Noxolo, Richard Baxter, Patricia Daley, and Margaret Byron. "The 2017 RGS-IBG Chair's Theme: Decolonising Geographical Knowledges, or Reproducing Coloniality?" *Area* 49, no. 3 (2017): 384–88.

Evans, Edward. *South with Scott*. London: Collins, 1921.

Evans, Sarah L. "Terra Incognita: Women on Royal Geographical Society Supported Expeditions, 1913–1970." PhD diss., University of the West of England, 2015.

———. "Terra Incognita: Women's Expeditions in the Arctic, c. 1913–1955." Presentation to the International Congress of Historical Geographers, July 19, 2018, Warsaw, Poland.

Evans, Sarah L., Innes M. Keighren, and Avril Maddrell. "Coming of Age? Reflection on the Centenary of Women's Admission to the Royal Geographical Society." *Geographical Journal* 179, no. 4 (2013): 373–76.

Fabian, Johannes. *Out of Our Minds: Reason and Madness in the Exploration of Central Africa*. Berkeley: University of California Press, 2000.

Felber, Lynette. *Clio's Daughters: British Women Making History, 1790–1899*. Newark: University of Delaware Press, 2007.

Fernandez-Armesto, Felipe. *Pathfinders: A Global History of Exploration*. New York: W. W. Norton, 2006.

Fiennes, Ranulph. *Captain Scott*. London: Hodder & Stoughton, 2011.

———. *Shackleton*. London: Penguin, 2021.

Finkelstein, David. *The House of Blackwood: Author-Publisher Relation in the Victorian Era*. University Park: Penn State University Press, 2002.

———. "Unraveling Speke: The Unknown Revision of an African Exploration Classic." *History in Africa* 30 (2003): 117–32.

Fiske, Shanyn. "Orientalism Reconsidered: China and the Chinese in Nineteenth-Century Literature and Victorian Studies." *Literature Compass* 8, no. 4 (2011): 214–26.

Fleetwood, Lachlan. "Bodies in High Places: Exploration, Altitude Sickness, and the Problem of Bodily Comparison in the Himalaya, 1800–50." *Itinerario* 43, no. 3 (2019): 489–515.

———. "No Former Travellers Having Attained Such a Height on the Earth's Surface: Instruments, Inscriptions, and Bodies in the Himalaya, 1800–1830." *History of Science* 56, no. 1 (2018): 3–34.

———. *Science on the Roof of the World: Empire and the Remaking of the Himalaya*. Cambridge: Cambridge University Press, 2021.

Foley, Tim. *Crean: The Extraordinary Life of an Irish Hero*. Stockport: Keel Foley, 2018.

Forsyth, Isla. "The More-than-human Geographies of Field Science." *Geography Compass* 7, no. 8 (2013): 527–39.

Forth, Christopher E. "Fat, Desire, and Disgust in the Colonial Imagination." *History Workshop Journal*, 73, no.1 (2012): 211–39.

———. *Masculinity in the Modern West: Gender Civilization and the Body*. Basingstoke: Palgrave Macmillan, 2008.

———. "Surviving our Paradoxes: Masculinity, Modernity, and the Body." *Culture, Society, and Masculinities* 1 (2010): 6–32.

Forth, Christopher E., and Anna Carden Coyne, eds. *Cultures of the Abdomen: Diet Digestion and Fat in the Modern World*. Basingstoke: Palgrave Macmillan, 2005.

Foster, Shirley and Sara Mills, eds. *An Anthology of Women's Travel Writing*. Manchester: Manchester University Press, 2002.

Foucault, Michel. *The Birth of the Clinic: An Archaeology of Medical Perception*. Routledge Classics ed. Translated by Alan M. Sheridan. London: Routledge, [1963] 2003.

———. *Discipline and Punish: The Birth of The Prison*. Translated by Allan Sheridan. New York: Pantheon, 1977.

———. *Power/Knowledge: Selected Interviews and Other Writings, 1972–1977*. Translated by Colin Gordon, Leo Marshall, John Mepham, and Kate Soper. New York: Pantheon, 1980.

———. *The Will to Knowledge: The History of Sexuality*. Vol. 1. Translated by Robert Hurley. London: Penguin, [1978] 1998.

Fox, William L. "Walking in Circles: Cognition and Science in High Places." In *High Places: Cultural Geographies of Mountains, Ice, and Science*, edited by Veronica della Dora and Denis Cosgrove, 19–32. London: I. B. Tauris, 2009.

Freshfield, Douglas and W. J. L. Wharton, eds. *Hints to Travellers: Scientific and General*. 7th ed. London: Royal Geographical Society 1893.

Galton, Francis. *The Art of Travel; or Shifts and Contrivances Available in Wild Countries*. 2nd ed. London: John Murray, 1856.

———. *Inquiries into Human Faculty and Its Development*. London: Macmillan, 1883.
Gappah, Petina. *Out of Darkness, Shining Light*. London: Faber & Faber, 2019.
Garltan, Luke. "'A Complete Craze': Isabella Bird Bishop in East Asia." *Photo-researcher*, no. 15 (2011): 13–26.
Ghosh, Durba. *Sex and the Family in Colonial India: The Making of Empire*. Cambridge: Cambridge University Press, 2006.
Gimlin, Debra. "What Is 'Body Work'? A Review of the Literature." *Sociology Compass* 1, no. 1 (2007): 353–70.
Godlewska, Anne Marie Claire. "From Enlightenment Vision to Modern Science? Humbolt's Visual Thinking." In *Geography and Enlightenment*, edited by Charles W. J. Withers and David N. Livingstone, 236–80. Chicago: University of Chicago Press, 1999.
Godsall, Jon R. "Fact and Fiction in Richard Burton's Personal Narrative of a Pilgrimage to El-Medinah and Meccah (1855–56)." *Journal of the Royal Asiatic Society* 3, no. 3 (1993): 331–51.
———. *The Tangled Web: A Life of Sir Richard Burton*. Leicester: Troubador, 2008.
Gopal, Priyamvada. *Insurgent Empire: Anticolonial Resistance and British Dissent*. London: Verso, 2020.
Graeber, David. *Debt: The First 5,000 Years*. New York: Melville House, 2011.
Graham Burnett, D. "'It Is Impossible to Make a Step without the Indians': Nineteenth-Century Geographical Exploration and the Amerindians of British Guiana." *Ethnohistory* 49, no. 1 (2002): 3–40.
Grant, James A. *A Walk across Africa: Domestic Scenes from My Nile Journal*. Edinburgh: Blackwood, 1864.
Green, Marcus E. "Rethinking the Subaltern and the Question of Censorship in Gramsci's Prison Notebooks." *Postcolonial Studies* 14, no. 4 (December 1, 2011): 387–404.
Gregory, Derek. *Geographical Imaginations*. Oxford: Blackwell, 1994.
Gregory, James. *Of Victorians and Vegetarians: The Vegetarian Movement in Nineteenth-Century Britain*. London: I. B. Tauris, 2007.
Grosz, Elisabeth. *Volatile Bodies: Towards a Corporeal Feminism*. Indianapolis: Indiana University Press, 1994.
Guasco, Anna. "On an Ethic of Not Going There." *Geographical Journal* 188, no. 3 (September 1, 2022): 468–75.
Guha, Ranajit. *Elementary Aspects of the Peasant Insurgency in Colonial India*. Durham NC: Duke University Press, [1983] 1999.
Guha, Ranajit, ed. *Subaltern Studies 1: Writings on Asian History and Society*. Delhi: Oxford University Press, 1982

———. *A Subaltern Studies Reader, 1986–1995*. Minneapolis: University of Minnesota Press, 1996.
Guly, Henry R. "Use and Abuse of Alcohol and Other Drugs during the Heroic Age of Antarctic Exploration." *History of Psychiatry* 24, no. 1 (2013): 94–105.
Gurney, Edmund, Frederic Myers, and Frank Podmore, *Phantasms of the Living*. London: Society for Psychical Research, 1886.
Hadwen, Ian. "The Literature of Burke and Wills." *Queensland History Journal* 21, no. 8 (2012): 537–50.
Hansard Parliamentary Debates. 3rd series (1830–91).
Haraway, Donna J. *The Companion Species Manifesto: Dogs, People, and Significant Otherness*. Chicago: Prickly Paradigm, 2006.
———. *When Species Meet*. Minneapolis: University of Minnesota Press, 2008.
Harris, Dawn P. *Punishing the Black Body: Marking Social and Racial Structures in Barbados and Jamaica*. Athens: University of Georgia Press, 2017.
Harrison, Mark. *Climates and Constitutions: Health, Race, Environment, and British Imperialism in India*. Oxford: Oxford University Press, 1999.
Hayes, Emily. "Geographical Light: The Magic Lantern, the Reform of the Royal Geographical Society, and the Professionalization of Geography, c. 1885–1894." *Journal of Historical Geography* 62 (2018): 24–36.
Heffernan, Michael. "'A Dream as Frail as Those of Ancient Time': The in-Credible Geographies of Timbuctoo." *Environment and Planning D: Society and Space* 19, no. 2 (2001): 203–25.
Heggie, Vanessa. "Blood, Race, and Indigenous Peoples in Twentieth-Century Extreme Physiology." *History and Philosophy of the Life Sciences* 41, no. 2 (2019): 26.
———. *Higher and Colder: A History of Extreme Physiology and Exploration*. Chicago: University of Chicago Press, 2019.
———. "Why Isn't Exploration a Science?" *Isis* 105, no. 2 (2014): 318–34.
Heilprin, Angelo. "The British National Antarctic Expedition." *Bulletin of the American Geographical Society* 38, no. 3 (1906): 177–81.
Herschel, John Frederick William. *A Manual of Scientific Enquiry*. London: John Murray, 1849.
Herzig, Rebecca. *Suffering for Science: Reason and Sacrifice in Modern America*. New Brunswick NJ: Rutgers University Press, 2005.
Hevly, Bruce. "The Heroic Science of Glacier Motion." *Osiris* 11, no. 1 (1996): 66–86.
Hill, Jen. *White Horizon: The Arctic in the Nineteenth-Century British Imagination*. New York: State University of New York Press, 2009.
Hirsch, Lioba. "Mental Health, Coloniality, and Fieldwork in the European University: A Reflection in Three Challenges—Commentary to Taylor."

Fennia—International Journal of Geography 198, nos. 1–2 (September 17, 2020): 210–13.

Hochschild, Adam. *King Leopold's Ghost: A Story of Greed, Terror, and Heroism in Colonial Africa*. Boston: Houghton Mifflin, 1999.

Hodder, Jake. "On Absence and Abundance: Biography as Method in Archival Research." *Area* 49, no. 4 (2017): 452–59.

Hokowhitu, Brendan, Aileen Moreton-Robinson, Linda Tuhiwai-Smith, Chris Andersen, and Steve Larkin. *Routledge Handbook of Critical Indigenous Studies*. New York: Routledge, 2020.

Holton, Mark. "On the Geographies of Hair: Exploring the Entangled Margins of the Bordered Body." *Progress in Human Geography* 44, no. 3 (2020): 555–71.

hooks, bell. *Black Looks: Race and Representation*. Boston: South End, 1992.

Hooper, Meredith. *The Longest Winter: Scott's Other Heroes*. London: John Murray, 2010.

Howell, Philip. *Geographies of Regulation: Policing Prostitution in Nineteenth-Century Britain and the Empire*. Cambridge: Cambridge University Press, 2009.

Howkins, Adrian. "'Have You Been There?' Some Thoughts on (Not) Visiting Antarctica." *Environmental History*, July 1, 2010: 514–19.

Howson, Alexandra. *Embodying Gender*. London: SAGE, 2005.

Hribal, Jason. "Animals Are Part of the Working Class: A Challenge to Labor History." *Labor History* 44, no. 4 (2003): 435–45.

Hsü, Immanuel C. Y. *The Rise of Modern China*. 6th ed. Oxford: Oxford University Press, 2000.

Humphries, John. *Search for the Nile's Sources: The Ruined Reputation of John Petherick, Nineteenth-Century Welsh Explorer*. Cardiff: Cardiff University Press, 2013.

Huntford, Roland. *The Last Place on Earth: Scott and Amundsen's Race to the South Pole*. New York: Modern Library, 1999.

Hussey, Kristen D. *Imperial Bodies in London: Empire, Mobility, and the Making of British Medicine, 1880–1914*. Pittsburgh PA: University of Pittsburgh Press, 2021.

———. "Imperial Mobilities: Tropical Bodies and the Spaces of Medical Practice and Research in London, c. 1880–1914." PhD diss., Queen Mary, University of London, 2018.

Huxley, Elspeth J. G. *Scott of the Antarctic*. Lincoln: University of Nebraska Press, 1990.

Hyam, Ronald. *Empire and Sexuality: The British Experience*. Manchester: Manchester University Press, 1990.

Iman Jackson, Zakiyyah. *Becoming Human: Matter and Meaning in an Antiblack World*. New York: New York University Press, 2020.

Ireland, Deborah. *Isabella Bird: A Photographic Journal of Travels through China, 1894–1896*. London: Royal Geographical Society, 2015.
Jackson, Frederick G. *The Great Frozen Land (Bolshaia Zemelskija Tundra): Narrative of a Winter Journey across the Tundras and a Sojourn among the Samoyads*. London: Macmillan, 1895.
———. *A Thousand Days in the Arctic*. 2 vols. London: Harper & Brothers, 1899.
Jäger, Gustav. *Selections from Essays on Health Culture: And the Sanitary Woollen System*. London: Jäger, 1884.
Jay, Martin. *Downcast Eyes: The Denigration of Vision in Twentieth-Century French Thought*. Berkeley: University of California Press, 1993.
Jazeel, Tariq. "Mainstreaming Geography's Decolonial Imperative." *Transactions of the Institute of British Geographers* 42, no. 3 (2017): 334–37.
———. "Subaltern Geographies: Geographical Knowledge and Postcolonial Strategy." *Singapore Journal of Tropical Geography* 35, no. 1 (2014): 88–103.
Jazeel, Tariq, and Stephen Legg. *Subaltern Geographies*. Athens: University of Georgia Press, 2019.
Jazeel, Tariq, Anson Mackay, Sarah L. Evans, and Catherine Souch. "Decolonising Spaces of Geographical Knowledge Production: 'Thinking Geographically' with Undergraduate Geographers in the RGS-IBG at Kensington Gore." *Area* 54, no. 3 (2022): 516–20.
Jeal, Tim. *Explorers of the Nile: The Triumph and Tragedy of a Victorian Adventure*. London: Faber & Faber, 2011.
———. *Livingstone*. Rev. ed. New Haven CT: Yale University Press, [1973] 2013.
———. *Stanley: The Impossible Life of Africa's Greatest Explorer*. London: Faber & Faber, 2007.
Johnson, Azeezat. "An Academic Witness: White Supremacy within and beyond Academia." In *The Fire Now: Anti-Racist Scholarship in Times of Explicit Racial Violence*, edited by Azeezat Johnson, Remi Joseph-Salisbury, and Beth Kamunge, 15–25. London: Zed, 2018.
———. "Throwing Our Bodies against the White Background of Academia." *Area* 52, no. 1 (2020): 89–96.
Johnson, Nuala C. "Illustrating Nature: Exploration, Natural History, and the Travels of Charlotte Wheeler-Cuffe in Burma." In *Geographies of Knowledge: Science, Scale, and Spatiality in the Nineteenth Century*, edited by Robert J. Mayhew and Charles W. J. Withers, 187–214. Baltimore MD: John Hopkins University Press, 2020.
———. "On the Colonial Frontier: Gender, Exploration and Plant-Hunting on Mount Victoria in Early Twentieth-Century Burma." *Transactions of the Institute of British Geographers* 42, no. 3 (2017): 417–31.

Johnston, Andrew. "'Arbitrary and Cruel Punishments': Trends in Royal Navy Courts Martial, 1860–1869." *International Journal of Maritime History* 33, no. 3 (August 1, 2021): 525–44.

Johnston, Catherine. "Beyond the Clearing: Towards a Dwelt Animal Geography." *Progress in Human Geography* 32, no. 5 (2008): 633–49.

Johnston, Harry Hamilton. *The Nile Quest: A Record of the Exploration of the Nile and Its Basins*. London: Lawrence & Bullen, 1903.

Jones, Aubrey A. *Scott's Forgotten Surgeon: Dr. Reginald Koettlitz, Polar Explorer*. Caithness: Whittles, 2011.

Jones, Cam Sharp. "Animals, Joseph Dalton Hooker, and the Ross Expedition to Antarctica." *Journal for Maritime Research* 22, nos. 1–2 (2020): 25–40.

Jones, Lowri. "Local Knowledge and Indigenous Agency in the History of Exploration: Studies From the RGS-IBG Collection." PhD diss., Royal Holloway University of London, 2010.

Jones, Max. "Exploration, Celebrity, and the Making of a Transnational Hero: Fridtjof Nansen and the Fram Expedition." *Journal of Modern History* 93, no. 1 (2021): 68–108.

———. "From 'Noble Example' to 'Potty Pioneer:' Rethinking Scott of the Antarctic, c. 1945–2011." *Polar Journal* 1, no. 2 (2011): 191–206.

———. *The Last Great Quest: Captain Scott's Antarctic Sacrifice*. Oxford: Oxford University Press, 2003.

———. "'The Truth about Captain Scott': *The Last Place on Earth*, Debunking, Sexuality, and Decline in the 1980s." *Journal of Imperial and Commonwealth History* 42, no. 5 (2014): 857–81.

———. "What Should Historians Do with Heroes? Reflections on Nineteenth- and Twentieth-Century Britain." *History Compass* 5, no. 2 (2007): 439–54.

Jones, Max, Berny Sèbe, Bertrand Taithe, and Peter Yeandle, eds. *Decolonising Imperial Heroes: Cultural Legacies of British and French Empires*. Abingdon: Routledge, 2016.

Jones, Max, Berny Sèbe, John Strachan, Bertrand Taithe, and Peter Yeandle. "Decolonising Imperial Heroes." *Journal of Imperial and Commonwealth History* 42, no. 5 (2014): 778–79.

Jones, Owain. "(Un)ethical Geographies of Human–Non-Human Relations: Encounters, Collectives, and Spaces." In *Animal Spaces, Beastly Places*, edited by Chris Philo and Chris Wilbert, 267–90. London: Routledge, 2000.

Kaalund, Nanna Katrine Lüders. "Erasure as a Tool of Nineteenth-Century European Exploration, and the Arctic Travels of Tookoolito and Ipiirvik." *Historical Journal*, August 17, 2022, 1–19.

———. *Explorations in the Icy North: How Travel Narratives Shaped Arctic Science in the Nineteenth Century*. Pittsburgh PA: University of Pittsburgh Press, 2021.

Kanasaka, Kiyonori, *Isabella Bird and Japan: A Reassessment*. Translated by Nicholas Pertwee. Folkstone: Renaissance, 2017.

Kato, Daniela. "'I Write the Truth as I See It:' Unsettling the Boundaries of Gender, Travel Writing, and Ethnography in Isabella Bird's *Unbeaten Tracks in Japan*.'" In *Women in Transit through Literary Liminal Spaces*, edited by Teresa Gómez Reus and Terry Gifford, 77–90. Basingstoke: Palgrave Macmillan, 2013.

Kearns, Gerry. *Geopolitics and Empire: The Legacy of Halford Mackinder*. Oxford: Oxford University Press, 2009.

———. "Topple the Racists 1: Decolonising the Space and Institutional Memory of the University." *Geography* 105, no. 3 (2020): 116–25.

———. "Topple the Racists 2: Decolonising the Space and the Institutional Memory of Geography." *Geography* 106, no. 1 (2021): 4–15.

Keighren, Innes M. "Reading the Reception of Ellen Churchill Semple's Influences of Geographic Environment (1911)." PhD diss., University of Edinburgh, 2008.

———. "'A Royal Geographical Society for Ladies': The Lyceum Club and Women's Geographical Frontiers in Edwardian London." *Professional Geographer* 69 (2017): 661–69.

Keighren, Innes M., Charles. W. J. Withers, and Bill Bell. *Travels into Print: Exploration, Writing, and Publishing with John Murray, 1773–1859*. Chicago: University of Chicago Press, 2015.

Kennedy, Dane. "'Captain Burton's Oriental Muck Heap': *The Book of the Thousand Nights* and the Uses of Orientalism." *Journal of British Studies* 39, no. 3 (2000): 317–39.

———. *The Highly Civilized Man: Richard Burton and the Victorian World*. Cambridge MA: Harvard University Press, 2005.

———. *Imperial History Wars: Debating the British Empire*. London: Bloomsbury, 2018.

———. *The Last Blank Spaces: Exploring Africa and Australia*. Cambridge MA: Harvard University Press, 2013.

———. "Lost in Place: Two Expeditions Gone Awry in Africa." In *Geographies of Knowledge*, edited by Robert Mayhew and Charles W. J. Withers, 235–54. Baltimore MD: John Hopkins University Press, 2020.

———. *The Magic Mountains: Hill Stations and the Raj*. Berkeley: University of California Press, 1996.

———. "The Perils of the Midday Sun: Climatic Anxieties in the Colonial Tropics." In *Imperialism and the Natural World*, edited by John M. Mackenzie, 118–40. Manchester: Manchester University Press, 1990.

Kennedy, Dane, ed. *Reinterpreting Exploration: The West in the World*. Oxford: Oxford University Press 2014.

Kim, Nanny. "River Control, Merchant Philanthropy, and Environmental Change in Nineteenth-Century China." *Journal of the Economic and Social History of the Orient* 52, nos. 4–5 (2009): 660–94.

Kingsley, George. *Notes on Sport and Travel*. London: Macmillan, 1900.

Kingsley, Mary. *Travels in West Africa, Congo Francais, Corisco, and Cameroons*. London: Macmillan, 1897.

Koettlitz, Reginald. "Notes on the Galla of Walega and the Bertat." *Journal of the Anthropological Institute of Great Britain and Ireland* 30 (1900): 50–55.

Konishi, Shino, Maria Nugent, and Tiffany Shellam, eds. *Indigenous Intermediaries: New Perspectives on Exploration Archives*. Acton: Australia National University Press, 2015.

Kristeva, Julia. *Powers of Horror: An Essay on Abjection*. Translated by Leon S. Roudiez. New York: Columbia University Press.

Laing, Stuart. *Tippu Tip: Ivory, Slavery, and Discovery in the Scramble for Africa*. Surbiton: Medina, 2017.

Lamb, Johnathan. *Scurvy: The Disease of Discovery*. Princeton NJ: Princeton University Press, 2017.

Lambert, David. *Mastering the Niger: James MacQueen's African Geography and the Struggle over Atlantic Slavery*. Chicago: University of Chicago Press, 2013.

Laqueur, Thomas W. *Making Sex: Body and Gender from the Greeks to Freud*. Cambridge MA: Harvard University Press, 1990.

Larson, Edward J. *An Empire of Ice: Scott, Shackleton, and the Heroic Age of Antarctic Science*. New Haven CT: Yale University Press, 2011.

———. *To the Edges of the Earth: 1909, the Race for the Three Poles, and the Climax of the Age of Exploration*. New York: Harper Collins, 2018.

Lawrence, Anya, and Natasha Dowey. "Six Simple Steps towards Making GEES Fieldwork More Accessible and Inclusive." *Area* 54, no. 1 (2022): 52–59.

Lawrence, Christopher, and Steven Shapin, eds. *Science Incarnate: Historical Embodiments of Natural Knowledge*. Chicago: University of Chicago Press, 1998.

Leake, Martin. "Is the Quorra, Which Has Lately Been Traced from its Discharge into the Sea, the Same River as the Nigir of the Ancients?" *Journal of the Royal Geographical Society* 2 (1832): 1–28.

Leane, Elizabeth. "Animals." In *The Routledge Research Companion to Travel Writing*, edited by Alasdair Pettinger and Tim Youngs, 305–18. London: Routledge, 2020.

Leduc-Grimaldi, Mathilde, and James L. Newman. *Finding Dr. Livingstone: A History in Documents from the Henry Morton Stanley Archives*. Athens: Ohio University Press, 2020.

Legg, Stephen. "Beyond the European Province: Foucault and Postcolonialism."

In *Space, Power, Knowledge*, edited by Jeremy W. Crampton and Stuart Elden, 265–90. London: Routledge, 2007.
———. "Decolonialism." *Transactions of the Institute of British Geographers* 42, no. 3 (2017): 345–48.
———. *Prostitution and the Ends of Empire: Scale Governmentalities and Interwar India*. Durham NC: Duke University Press, 2014.
Legg, Stephen, and Deana Heath, ed. *South Asian Governmentalities: Michel Foucault and the Question of Postcolonial Orderings*. Cambridge: Cambridge University Press, 2018.
Lemtur, Nokmedemla. *Locating Himalayan Porters in the Archivalien Der Expeditionsgesellschaften of the German Alpine Club (1929–1939)*. Delhi: Max Weber Forum for South Asian Studies, 2020.
Leshem, Noam, and Alasdair Pinkerton. "Rethinking Expeditions: On Critical Expeditionary Practice." *Progress in Human Geography* 43, no. 3 (June 1, 2019): 496–514.
Lewis-Jones, Huw. *Face to Face: Polar Portraits*. Cambridge: Scott Polar Research Institute, 2008.
Lewis, Joanna. *Empire of Sentiment: The Death of Livingstone and the Myth of Victorian Imperialism*. Cambridge: Cambridge University Press, 2018.
Leyland, Julian, Hilary Geoghegan, Sarah Marie Hall, Alan Latham, and Catherine Souch. "Classics Revisited: 'Muddy Glee'—What Geography Fieldwork Means in the Current Moment." *Area* 54, no. 4 (2022): 522–24.
Lindsay, David. *Journal of the Elder Exploring Expedition, 1891*. Adelaide: Government Printer, 1893.
Livingstone, David N. *Adam's Ancestors: Race, Religion, and the Politics of Human Origins*. Baltimore MD: John Hopkins University Press, 2008.
———. *The Geographical Tradition: Episodes in the History of a Contested Enterprise*. Oxford: Blackwell, 1992.
———. *The Last Journals of David Livingstone, in Central Africa*. 2 vols. Edited by Horace Waller. London: John Murray, 1874.
———. *Missionary Travels and Researches in South Africa*. London: John Murray, 1857.
———. "Politics, Culture, and Human Origins: Geographies of Reading and Reputation in Nineteenth-Century Science." In *Geographies of Nineteenth Century Science*, edited by David N. Livingstone and Charles W. J. Withers, 178–202. Chicago: University of Chicago Press, 2011.
———. *Putting Science in Its Place*. Chicago: University of Chicago Press, 2003.
———. "Race, Space, and Moral Climatology: Notes towards a Genealogy." *Journal of Historical Geography* 28, no. 2 (2002): 159–80.

Livingstone, David. N. and Charles. W. J. Withers, eds. *Geography and Enlightenment*. Chicago: University of Chicago Press, 1999.

Livingstone, Justin D. "A 'Body' of Evidence: The Posthumous Presentation of David Livingstone." *Victorian Literature and Culture* 40, no. 1 (2012): 1–24.

———. *Livingstone's "Lives": A Metabiography of a Victorian Icon*. Manchester: Manchester University Press, 2014.

———. "A Romance of Slavery: Exploration, Encounters, and Cartographies of Violence in H. M. Stanley's *My Kalulu*." *Studies in Travel Writing* 21, no. 4 (October 2, 2017): 349–68.

Longhurst, Robyn. *Bodies: Exploring Fluid Boundaries*. London: Routledge, 2000.

Lorimer, Hayden, and Nick Spedding. "Locating Field Science: A Geographical Family Expedition to Glen Roy, Scotland." *British Journal for the History of Science* 38, no. 1 (2005): 13–33.

MacDonald, Fraser, and Charles W. J. Withers, eds. *Geography, Technology, and Instruments of Exploration*. Farnham: Ashgate 2015.

MacGregor, Arthur, ed. *Naturalists in the Field: Collecting, Recording, and Preserving the Natural World from the Fifteenth to the Twenty-First Century*. Leiden: Brill, 2018.

Mackay, Anson W., David Adger, Alexander L. Bond, Sam Giles, and Erinma Ochu. "Straight-Washing Ecological Legacies." *Nature Ecology & Evolution* 3, no. 12 (December 2019): 1611.

MacKenzie, John M. "David Livingstone and the Worldly After-Life: Imperialism and Nationalism in Africa." In *David Livingstone and the Victorian Encounter with Africa*, edited by John M. Mackenzie and Joanna Skipwith, 201–19. London: National Portrait Gallery, 1996.

———. "David Livingstone—Prophet or Patron Saint of Imperialism in Africa: Myths and Misconceptions." *Scottish Geographical Journal* 129, nos. 3–4 (December 1, 2013): 277–91.

———. *The Empire of Nature: Hunting, Conservation, and British Imperialism*. Manchester: Manchester University Press, 1997.

———. "The Iconography of Exemplary Life: The Case of David Livingstone." In *Heroic Reputations and Exemplary Lives*, edited by Geoffrey Cubbitt and Alan Warren, 84–104. Manchester: Manchester University Press, 2000.

Maddison, Ben. *Class and Colonialism in Antarctic Exploration, 1750–1920*. London: Pickering & Chatto, 2014.

Maddrell, Avril. *Complex Locations: Women's Geographical Work in the UK, 1850–1970*. Oxford: Wiley-Blackwell 2009.

Maguire, Sarah. "Gender Differences in Attitudes to Undergraduate Fieldwork." *Area* 30, no. 3 (1998): 207–14.

Maitland, Alexander. *Speke and the Discovery of the Source of the Nile*. Trowbridge: Readers Union, 1973.

Markham, Albert Hastings. *The Great Frozen Sea: A Personal Narrative of the Voyage of the* Alert *during the Arctic Expedition of 1875–76*. London: Daldy Ibister, 1878.

Markham, Clements. "Antarctic Disaster." *Geographical Journal* 41, no. 3 (March 1913): 201–22.

———. *Antarctic Obsession: A Personal Narrative of the Origins of the British National Antarctic Expedition, 1901–4*. Harleston: Bluntisham, 1986.

———. "The First Year's Work of the National Antarctic Expedition." *Geographical Journal* 22, no. 1 (July 1903): 13–20, 37.

Martin, Emily. "The Tintometer, Anthropology, and the Science of Color." *Anthropology Now* 8, no. 2 (2016): 14–24.

Martin, Peter R. "The 'Deer-Men' and the 'Bowhead-Men': The Colonial Cooptation of Arctic Indigenous Knowledge within the 'Origins of the Inuit' Debates." *Transactions of the Institute of British Geographers* 49, no. 1 (2024): e12615.

———. "Indigenous Tales of the Beaufort Sea: Arctic Exploration and the Circulation of Geographical Knowledge." *Journal of Historical Geography* 67 (2020): 24–35.

May, Karen. "Could Captain Scott Have Been Saved? Cecil Meares and the 'Second Journey' That Failed: A Response to Bill Alp." *Polar Record* 54, no. 2 (2018): 183–87.

May, Karen, and George Lewis. "'Strict Injunctions That the Dogs Should Not Be Risked': A Revised Hypothesis for This Anecdote and Others in Narratives of Scott's Last Expedition." *Polar Record* 55, no. 6 (2019): 1–12.

———. "'They Are Not the Ponies They Ought to Have Been': Revisiting Cecil Meares' Purchase of Siberian Ponies for Captain Scott's British Antarctic (Terra Nova) Expedition (1910–1913)." *Polar Record* 51, no. 6 (2015): 655–66.

Mayhew, Robert. "Halford Mackinder's 'New' Political Geography and the Geographical Tradition." *Political Geography* 19, no. 6 (August 1, 2000): 771–91.

Mays, S., and O. Beattie. "Evidence for End-Stage Cannibalism on Sir John Franklin's Last Expedition to the Arctic, 1845." *International Journal of Osteoarchaeology* 26, no. 5 (2016): 778–86.

McClintock, Anne. *Imperial Leather: Race Sexuality and the Colonial Contest*. London: Routledge, 1995.

McClintock, Leopold. "Arctic Sledge Travelling." In *The Antarctic Manual: For the Use of the Expedition of 1901*, edited by George Murray, 293–304. London: Royal Geographical Society, 1901.

McCook, Stuart. "'It May Be Truth, But It Is Not Evidence': Paul Du Chaillu and the Legitimation of Evidence in the Field Sciences." *Osiris* 11 (1996): 177–97.
McCorristine, Shane. *The Spectral Arctic: A History of Dreams and Ghosts in Polar Exploration*. London: UCL Press, 2018.
McCorristine, Shane, and Jane S. P. Mocellin. "Christmas at the Poles: Emotions, Food, and Festivities on Polar Expeditions, 1818–1912." *Polar Record* 52, no. 5 (2016): 562–77.
McDowell, Linda. *Gender Place, and Identity: Understanding Feminist Geographies*. Minnesota: University of Minneapolis Press, 1999.
———. *Working Bodies: Interactive Service Employment and Workplace Identities*. Oxford: Wiley-Blackwell, 2009.
McLynn, Frank. *Stanley: Sorcerer's Apprentice*. Oxford: Oxford University Press, 1992.
McRae, Calista. "'The Great Chain of Being Come Undone': Linking Blackness and Animal Studies." *Environmental Humanities* 14, no. 1 (March 1, 2022): 202–15.
Merleau-Ponty, Maurice. *Phenomenology of Perception*. Translated by Donald A. Landes. Abingdon: Routledge, [1945] 2012.
———. *The Visible and the Invisible*. Edited by Claude Lefort. Translated by Alphonso Lingis. Evanston IL: Northwestern University Press, 1968.
Michael, Boris. "Making Mount Kilimanjaro German: Nation Building and Heroic Masculinity in the Colonial Geographies of Hans Meyer." *Transactions of the Institute of British Geographers* 44, no. 3 (2019): 493–508.
Middleton, Dorothy. "The Search for the Nile Sources." *Geographical Journal* 138, no. 2 (1972): 209–21.
———. "Some Victorian Lady Travellers." *Geographical Journal* 139, no. 1 (1973): 65–75.
———. *Victorian Lady Travellers*. Chicago: Academy, 1982.
Mills, Sarah. *Discourses of Difference: An Analysis of Women's Travel Writing and Colonialism*. London: Routledge, 1991.
Montefiore, Arthur. "The Jackson-Harmsworth North Polar Expedition: An Account of Its First Winter and of Some Discoveries in Franz Josef Land." *Geographical Journal* 6, no. 6 (1895): 499–519.
Moore, Fiona. "Comparing Colonialisms in Dan Simmons' Novel *The Terror* and Its AMC Adaptation." 2019. https://www.academia.edu/40129260/Comparing_colonialisms_in_Dan_Simmons_novel_The_Terror_and_its_amc_adaptation (accessed September 19, 2023).
Moorehead, Alan. *The White Nile*. London: Hamish & Hamilton, 1960.
Morell Margot, and Capparell, Stephanie. *Shackleton's Way: Leadership Lessons from the Great Antarctic Explorer*. London: John Murray, 2011.

Morin, Karen M. "Carceral Space: Prisoners and Animals." *Antipode* 48, no. 5 (2016): 1317–36.

———. *Carceral Space: Prisoners and Animals*. New York: Routledge, 2018.

———. *Frontiers of Femininity: A New Historical Geography of the Nineteenth-Century American West*. Ithaca NY: Syracuse University Press, 2008.

Morris, Pat. "Reflections on Some Practical Aspects of Collecting during the Nineteenth and Twentieth Centuries." In *Naturalists in the Field*, edited by Arthur MacGregor, 759–74. Leiden: Brill, 2018.

Morris, Richard E. "The Victorian 'Change of Air' as Medical and Social Construction." *Journal of Tourism History* 10, no. 1 (2018): 49–65.

Mossakowski, Tomasz Filip. "'The Sailors Dearly Love to Make Up': Cross-Dressing and Blackface during Polar Exploration." PhD diss., King's College London, 2014.

Mueggler, Eric. *The Paper Road: Archive and Experience in the Botanical Exploration of West China and Tibet*. Berkeley: University of California Press, 2011.

Murray, Carl. "The Use and Abuse of Dogs on Scott's and Amundsen's South Pole Expeditions." *Polar Record* 44, no. 4 (2008): 303–10.

Murray, T. Douglas, and Arthur Silva White. *Sir Samuel Baker; a Memoir*. London: Macmillan, 1895.

Nagai, Kaori. "Vermin Writing." *Journal for Maritime Research* 22, nos. 1–2 (July 2, 2020): 59–74.

Nares, George. *Narrative of a Voyage to the Polar Sea during 1875–76 in H.M. Ships "Alert" and "Discovery."* 2 vols. London: Sampson Low, Marston, Searle, & Rivington, 1878.

Naylor, Simon, and James R. Ryan, eds. *New Spaces of Exploration: Geographies of Discovery in the Twentieth Century*. London: I. B. Tauris, 2010.

Nelson, Lise. "Bodies (and Spaces) Do Matter: The Limits of Performativity." *Gender, Place & Culture* 6, no. 4 (December 1999): 331–53.

Newman, L. James. *Paths without Glory: Richard Francis Burton in Africa*. Washington DC: Potomac, 2010.

Nicolaïdis, Kalypso, Berny Sèbe, and Gabrielle Maas, eds. *Echoes of Empire: Memory, Identity and Colonial Legacies*. London: I. B. Tauris, 2014.

North, Michelle A., Warwick W. Hastie, and Lauren Hoyer. "Out of Africa: The Underrepresentation of African Authors in High-Impact Geoscience Literature." *Earth-Science Reviews* 208 (September 1, 2020): 103262.

Nourse, J. E. *Narrative of the Second Arctic Expedition Made by Charles F. Hall: His Voyage to Repulse Bay, Sledge Journeys to the Straits of Fury and Hecla and to King William's Land, and Residence among the Eskimos, during the Years 1864-'69*. Washington DC: Government Printing Office, 1879.

Noxolo, Patricia. "Decolonial Theory in a Time of the Re-colonisation of UK Research." *Transactions of the Institute of British Geographers* 42, no. 3 (2017): 342–44.

———. Interview with GeogPod, Geographical Association Podcast, November 26, 2020. https://www.geography.org.uk/GeogPod-The-gas-Podcast.

———. "Introduction: Decolonising Geographical Knowledge in a Colonised and Re-colonising Postcolonial World." *Area* 49, no. 3 (2017): 317–19.

O'Hanlon, Rosalind. "Recovering the Subject Subaltern Studies and Histories of Resistance in Colonial South Asia." *Modern Asian Studies* 22, no. 1 (1988): 189–224.

Oakley, Ann. *The Sociology of Housework*. Bristol: Policy, [1974] 2018.

Oates, Frank. *Matabele Land and the Victoria Falls: A Naturalist's Wanderings in the Interior of South Africa, from the Letters and Journals of the Late Frank Oates*. Edited by C.G. Oates. London: C. Kegan Paul, 1881.

Oliver, Catherine. *Animals of the Royal Geographical Society (with IBG): A Look into Wiley Digital Archives*. Hoboken NJ: Wiley, 2022.

———. "Beyond-Human Research: Negotiating Silence, Anger, and Failure in Multispecies Worlds." *Emotion, Space, and Society* 35 (May 1, 2020): 100686.

———. "Labour, Collaboration, Conflict, and Discovery: Animals of the RGS, Fieldnotes IV." 2021. https://catherinecmoliver.com/2021/07/14/labour-collaboration-conflict-and-discovery-finding-animals-in-the-royal-geographical-society-archives-animals-of-the-rgs-fieldnotes-iv/ (accessed February 19, 2024).

———. *Veganism, Archives, and Animals: Geographies of a Multispecies World*. Abingdon: Routledge, 2021.

Oreskes, Naomi. "Objectivity or Heroism? On the Invisibility of Women in Science." *Osiris* 11, no. 1 (1996): 87–113.

Otter, Chris. *The Victorian Eye: A Political History of Light and Vision in Britain, 1800–1910*. Chicago: University of Chicago Press, 2008.

Outram, Dorinda. "On Being Perseus: New Knowledge, Dislocation, and Enlightenment Exploration." In *Geography and Enlightenment*, edited by David N. Livingstone and Charles W. J. Withers, 281–94. Chicago: University of Chicago Press, 1999.

Palmer, Stephanie. "Emily Katharine Bates and the Suppression of Travail in Late Nineteenth-Century Travel Writing about the United States." *Studies in Travel Writing*, 14, no. 1 (2010): 29–42.

Pang, Alex Soojung-Kim. "Gender, Culture, and Astrophysical Fieldwork: Elizabeth Campbell and the Lick Observatory–Crocker Eclipse Expeditions." *Osiris* 11, no. 1 (1996): 17–43.

Park, Jihang. "Land of the Morning Calm, Land of the Rising Sun: The East Asian Travel Writings of Isabella Bird and George Curzon." *Modern Asian Studies* 36, no. 3 (2002): 513–34.

Park, Joohyun Jade. "Missing Link Found, 1880: The Rhetoric of Colonial Progress in Isabella Bird's Unbeaten Tracks in Japan." *Victorian Literature and Culture* 43, no. 2 (2015): 371–88.

Paterson, Mark. "Haptic Geographies: Ethnography, Haptic Knowledges, and Sensuous Dispositions." *Progress in Human Geography* 33, no. 6 (2009): 766–88.

Patterson, J. E. "Scott's Last Expedition." *Bookman* 45, no. 267 (December 1913): 66–67.

Pettitt, Clare. *Dr. Livingstone I Presume: Missionaries, Journalists, Explorers, and Empire*. London: Profile, 2013.

———. "Exploration in Print: From the Miscellany to the Newspaper." In *Reinterpreting Exploration: The West in the World*, edited by Dane Kennedy, 80–108. Oxford: Oxford University Press, 2014.

Phillips, Richard. "Writing Travel and Mapping Sexuality: Richard Burton's Sotadic Zone." In *Writes of Passage: Reading Travel Writing*, edited by James Duncan and Derek Gregory, 70–91. London: Routledge, 1999.

Philo, Chris, and Chris Wilbert, eds. *Animal Spaces, Beastly Places*. London: Routledge, 2000.

Phoenix, David Gary. "'More Like a Picnic Party': Burke and Wills: An Analysis of the Victorian Exploring Expedition of 1860–1861." PhD diss., James Cook University, 2018.

Pick, Daniel. *Faces of Degeneration: A European Disorder, c. 1848–1918*. Cambridge: Cambridge University Press, 1989.

Pickman, Sarah. "Dress, Image, and Cultural Encounter in the Heroic Age of Polar Exploration." In *Expedition: Fashion from the Extreme*, edited by Patricia Mears, 31–56. London: Thames & Hudson, 2017.

Pierce, Steven, and Anupama Rao. "Discipline and the Other Body: Humanism, Violence, and the Colonial Exception." In *Discipline and the Other Body: Correction, Corporeality, Colonialism*, edited by Steven Pierce and Anupama Rao, 1–35. Durham NC: Duke University Press, 2006.

Ponting, Herbert. *The Great White South; Being an Account of Experiences with Captain Scott's South Pole Expedition and of the Nature Life of the Antarctic*. London: Duckworth, 1922.

Porter, Roy. "History of the Body." In *New Perspectives on Historical Writing*, edited by Peter Burke, 206–32. University Park: Penn State University Press, 1991.

Poskett, James. *Materials of the Mind: Phrenology, Race, and the Global History of Science, 1815–1920*. Chicago: University of Chicago Press, 2019.

Potter, Russell A., Regina Koellner, Peter Carney, and Mary Williamson. *May We Be Spared to Meet on Earth: Letters of the Lost Franklin Arctic Expedition*. Montreal: McGill-Queen's University Press, 2022.

Pratt, Mary Louise. *Imperial Eyes: Travel Writing and Transculturation*. 2nd ed. London: Routledge, 2008.

Preston, Diana. *The Boxer Rebellion: The Dramatic Story of China's War on Foreigners that Shook the World in the Summer of 1900*. New York: Walker, 2000.

Priestley, Raymond E. *Antarctic Adventure: Scott's Northern Party*. New York: E. P. Dutton, 1915.

Puttick, Steve, and Amber Murrey. "Confronting the Deafening Silence on Race in Geography Education in England: Learning from Anti-Racist, Decolonial, and Black Geographies." *Geography* 105, no. 3 (November 1, 2020): 126–34.

Qiu, Qiang, Guojie Zhang, Tao Ma, Wubin Qian, Junyi Wang, Zhiqiang Ye, Changchang Cao, et al. "The Yak Genome and Adaptation to Life at High Altitude." *Nature Genetics* 44, no. 8 (2012): 946–49.

Quality Assurance Agency for Higher Education. "Subject Benchmark Statement: Geography." 2022. https://www.qaa.ac.uk/the-quality-code/subject-benchmark-statements/geography (accessed February 19, 2024).

Qureshi, Sadiah. *Peoples on Parade: Exhibitions, Empire, and Anthropology in Nineteenth-Century Britain*. Chicago: University of Chicago Press, 2011.

Radcliffe, Sarah A. "Decolonising Geographical Knowledges." *Transactions of the Institute of British Geographers* 42, no. 3 (2017): 329–33.

———. *Decolonizing Geography: An Introduction*. Cambridge: Polity, 2022.

Radcliffe, Sarah A., ed. "Decolonising Geographical Knowledges." Themed intervention, *Transactions of the Institute of British Geographers* 42, no. 3 (2017): 329–33.

Raj, Kapil. *Relocating Modern Science: Circulation and the Construction of Knowledge in South Asia and Europe, 1650–1900*. Basingstoke: Palgrave Macmillan, 2007.

———. "When Human Travellers Become Instruments: The Indo-British Exploration of Central Asia in the Nineteenth Century." In *Instruments, Travel, and Science: Itineraries of Precision from the Seventeenth to the Twentieth Century*, edited by Marie-Nöelle Bourguet, Christian Licoppe and H. Otto Sibum, 156–89. London: Routledge, 2002.

Rasmussen, Knud. "The Fifth Thule Expedition, 1921–24: The Danish Ethnographical and Geographical Expedition from Greenland to the Pacific." *Geographical Journal* 67, no. 2 (1926): 123–38.

———. *The Netsilik Eskimos: Social Life and Spiritual Culture*. Copenhagen: Gyldendal, 1931.

Rees, Jasper. *Blizzard: Race to the Pole*. London: BBC, 2006.

Rees, W. G. “Polar Mirages.” *Polar Record* 14, no. 150 (1988): 193–98.
Reeves, Edward A., ed. *Hints to Travellers: Scientific and General*. 2 vols. 9th ed. London: Royal Geographical Society, 1906.
———. “Surveying and Astronomical Observations.” In *Hints to Travellers: Scientific and General*. 9th ed. Edited by Edward A. Reeves, 1:1–12. London: Royal Geographical Society, 1906.
Regard, Frédéric. *British Narratives of Exploration: Case Studies on the Self and Other*. London: Pickering & Chatto, 2009.
———. “Fieldwork as Self-Harrowing: Richard Burton’s Cultural Evolution.” In *British Narratives of Exploration*, edited by Frederic Regard, 179–92. London: Pickering & Chatto, 2009.
Reid, Richard. “The Ganda on Lake Victoria: A Nineteenth-Century East African Imperialism.” *Journal of African History* 39, no. 3 (1998): 349–63.
———. “Images of an African Ruler: Kabaka Mutesa of Buganda, ca. 1857–1884.” *History in Africa* 26 (1999): 269–98.
Reidy, Michael S. “Mountaineering, Masculinity, and the Male Body in Mid-Victorian Britain.” *Osiris* 30 (2015): 158–81.
Rempel, Ruth. “‘No Better than a Slave or Outcast:’ Skill, Identity, and Power among the Porters of the Emin Pasha Relief Expedition, 1887–1890.” *International Journal of African Historical Studies* 43, no. 2 (2010): 279–318.
Report of the Admiralty Committee on Causes of the Outbreak of Scurvy in the Recent Arctic Expedition. London: Her Majesty’s Stationary Office, 1877.
Richards, Alfred Bate. *A Sketch of the Career of Richard F. Burton*. London: Waterlow & Sons, 1886.
Riffenburgh, Beau. *The Myth of the Explorer: The Press, Sensationalism, and Geographical Discovery*. London: Belhaven, 1994.
———. *Shackleton’s Forgotten Expedition: The Voyage of the Nimrod*. London: Bloomsbury, 2005.
Robb, George. “The Way of All Flesh: Degeneration, Eugenics, and the Gospel of Free Love.” *Journal of the History of Sexuality* 6, no. 4 (1996): 589–603.
Robinson, Michael F. *The Lost White Tribe: Explorers, Scientists, and the Theory That Changed a Continent*. Oxford: Oxford University Press, 2016.
———. “Manliness and Exploration: The Discovery of the North Pole.” *Osiris* 30, no. 1 (2015): 89–109.
———. “Scientific Travel.” In *The Cambridge History of Travel Writing*, edited by Nandini Das and Tim Youngs, 488–503. Cambridge: Cambridge University Press, 2019.
———. “Shackleton Syndrome.” *Isis* 11, no. 1 (2020): 112–19.
Rockel, Stephen J. “Caravan Porters of the Nyika: Labour, Culture, and Society in Nineteenth Century Tanzania.” PhD diss., University of Toronto, 1997.

———. *Carriers of Culture: Labor on the Road in Nineteenth-Century East Africa.* Portsmouth NH: Heinemann, 2006.

———. "Decentring Exploration in East Africa." In *Reinterpreting Exploration: The West in the World*, edited by Dane Kennedy,172–94. Oxford: Oxford University Press, 2014.

———. "Enterprising Partners: Caravan Women in Nineteenth-Century Tanzania." *Canadian Journal of African Studies/La Revie Canadienne des Etudes Africaines* 34, no. 3767 (2000): 748–78.

———. "'A Nation of Porters': The Nyamwezi and the Labour Market in Nineteenth-Century Tanzania." *Journal of African History* 41, no. 2 (2000): 173–95.

Rodaway, Paul. *Sensuous Geographies: Body, Sense, and Place*. Abingdon: Routledge, 1994.

Rogers, Steven L., Lisa Lau, Natasha Dowey, Hinna Sheikh, and Rebecca Williams. "Geology Uprooted! Decolonising the Curriculum for Geologists." *Geoscience Communication* 5, no. 3 (July 8, 2022): 189–204.

Rose, Gillian. *Feminism and Geography: The Limits of Geographical Knowledge*. Cambridge: Polity, 1993.

———. "Geography and Gender, Cartographies and Corporealities." *Progress in Human Geography* 19, no. 4 (1995): 544–48.

———. "On the Need to Ask: How, Exactly, Is Geography Visual?" *Antipode* 35, no. 2 (2003): 212–21.

Rowbotham, Judith. "'Soldiers of Christ?' Images of Female Missionaries in Late Nineteenth-Century Britain: Issues of Heroism and Martyrdom." *Gender & History* 12, no. 1 (2000): 82–106.

Royal Geographical Society (with IBG). "Council of Heads of Geography in UK Higher Education Institutions, Principles for Undergraduate Fieldwork." Version 1. December 2020. https://www.rgs.org/research/higher-education-resources/principles-for-undergraduate-field-courses (accessed April 24, 2024).

———. "Hidden Histories of Exploration." Accessed September 12, 2023. https://www.rgs.org/about/our-collections/online-exhibitions/hidden-histories-of-exploration/.

Royds, T. Roger. *The Diary of Lieutenant Charles W. R. Royds' RN Expedition to the Antarctic, 1901–1904*. Braidwood: self-published, 2001.

Ruberg, Willemijn. *History of the Body*. London: Springer, 2020.

Rupke, Nicolaas A. *Alexander Von Humboldt: A Metabiography*. Chicago: University of Chicago Press, 2008.

Ryan, James R. *Photography and Exploration*. London: Reaktion, 2013.

Saha, Jonathan. *Colonizing Animals: Interspecies Empire in Myanmar*. Cambridge: Cambridge University Press, 2021.

———. "Colonizing Elephants: Animal Agency, Undead Capital, and Imperial Science in British Burma." *BJHS Themes* 2 (2017): 169–89.

Said, Edward W. *Orientalism*. London: Penguin, [1978] 2003.

———. *Culture and Imperialism*. London: Vintage, 1994.

Salisbury, Laura, and Andrew Shail, eds. *Neurology and Modernity: A Cultural History of Nervous Systems: 1800–1959*. Basingstoke: Palgrave Macmillan, 2010.

Saunders, Edward. "Defining Metabiography in Historical Perspective: Between Biomyths and Documentary." *Biography* 38, no. 3 (2015): 325–42.

Scawen Blunt, Wilfrid. *My Diaries: Being a Personal Narrative of Events, 1888–1914*. London: Martin Secker, 1932.

Schaffer, Simon. "Easily Cracked: Scientific Instruments in States of Disrepair." *Isis* 102, no. 4 (2011): 706–17.

Schaffer, Simon, Lissa Roberts, Kapil Raj, and James Delbourgo, eds. *The Brokered World: Go-Betweens and Global Intelligence, 1770–1820*. Sagamore Beach MA: Science History, 2009.

Schelhaas, Bruno, Federico Ferretti, André Reyes Novaes, and Marcella Schmidt di Friedberg. *Decolonising and Internationalising Geography: Essays in the History of Contested Science*. London: Springer Nature, 2020.

Schoenbauer Thurin, Susan. *Victorian Travelers and the Opening of China, 1842–1907*. Athens: Ohio University Press, 1999.

Scott, Patrick. "Body-Building and Empire-Building: George Douglas Brown, the South African War, and *Sandow's Magazine of Physical Culture*." *Victorian Periodicals Review* 41, no. 1 (2008): 78–94.

Scott, Robert F. *Journals: Captain Scott's Last Expedition*. With an introduction by Max Jones. Oxford: Oxford University Press, 2006.

———. *Scott's Last Expedition*. 2 vols. Edited by Leonard Huxley. London: Smith & Elder, 1913.

———. *The Voyage of the* Discovery. 2 vols. London: Smith & Elder, 1905.

Sears, Jason. "Discipline in the Royal Navy, 1913–1946." *War & Society* 9, no. 2 (1991): 39–60.

Sèbe, Berny. *Heroic Imperialists in Africa: The Promotion of British and French Colonial Heroes, 1870–1939*. Manchester: Manchester University Press, 2013.

———. "The Making of British and French Legends of Exploration, 1821–1914." In *Reinterpreting Exploration: The West in the World*, edited by Dane Kennedy, 109–34. Oxford: Oxford University Press, 2014.

Seligmann, Mathew S. *Rum, Sodomy, Prayers, and the Lash Revisited: Winston Churchill and Social Reform in the Royal Navy, 1900–1915*. Oxford: Oxford University Press, 2018.

Shackleton, Ernest H. *Heart of the Antarctic: The Farthest South Expedition*. 2 vols. London: Heineman, 1909.
Shapin, Steven. "The Invisible Technician." *American Scientists* 77, no. 6 (1989): 554–63.
———. *Never Pure: Historical Studies of Science as if It Was Produced by People with Bodies, Situated in Time, Space, Culture, and Society, and Struggling for Credibility and Authority*. Baltimore MD: John Hopkins University Press, 2010.
———. "Placing the View from Nowhere: Historical and Sociological Problems in the Location of Science." *Transactions of the Institute of British Geographers* 23, no. 1 (1998): 5–12.
———. *A Social History of Truth: Science and Civility in Seventeenth-Century England*. Chicago: University of Chicago Press, 1994.
Sheldon, Mary French. *Sultan to Sultan: Adventures amongst the Masai and Other Tribes of East Africa*. Boston: Area, 1892.
Shellam, Tiffany. "Mediating Encounters through Bodies and Talk." In *Indigenous Intermediaries: New Perspectives on Exploration Archives*, edited by Shino Konishi, Maria Nugent, and Tiffany Shellam, 85–102. Acton: Australian National University Press, 2015.
Shellam, Tiffany, Maria Nugent, Shino Konishi, and Allison Cadzow, eds. *Brokers and Boundaries. Colonial Exploration in Indigenous Territory*. Acton: Australian National University Press, 2016.
Shilling, Chris. *The Body and Social Theory*. 2nd ed. London: SAGE, 2003.
Shipman, Pat. *To the Heart of the Nile: Lady Florence Baker and the Exploration of Central Africa*. London: William Morrow, 2004.
Simpson, Donald. *Dark Companions: The African Contribution to the European Exploration of East Africa*. London: Paul Elek, 1975.
Simpson, Thomas. *The Frontier in British India: Space, Science, and Power in the Nineteenth Century*. Cambridge: Cambridge University Press, 2021.
Sims, Richard. *Japanese Political History since the Meiji Renovation, 1868–2000*. Basingstoke: Palgrave Macmillan, 2001
Sinha, Mrinalini. *Colonial Masculinity: "The Manly Englishman" and the "Effeminate Bengali" in the Late Nineteenth Century*. Manchester: Manchester University Press, 1995.
Sivasundaram, Sujit. "The Human, the Animal, and the Prehistory of COVID-19." *Past & Present* 249, no. 1 (2020): 295–316.
———. "Imperial Transgressions: The Animal and the Human in the Idea of Race." *Comparative Studies of South Asia, Africa, and the Middle East* 35, no. 1 (2015): 156–72.
Skabelund, Aaron. "Animals and Imperialism: Recent Historiographical Trends." *History Compass* 11, no. 10 (2013): 801–7.

Skotnes-Brown, Jules. "Scurrying Seafarers: Shipboard Rats, Plague, and the Land/Sea Border." *Journal of Global History* 18, no. 1 (March 2023): 108–30.

Smith, Bernard. *European Vision and the South Pacific*. New Haven CT: Yale University Press, 1985.

Smith, Michael. *An Unsung Hero: Tom Crean—Antarctic Survivor*. Cork: Collins, 2000.

Solomon, Susan. *The Coldest March: Scott's Fatal Antarctic Expedition*. New Haven CT: Yale University Press, 2001.

Speke, John H. "Captain Speke's Discovery of the Victoria Nyanza Lake, the Supposed Source of the Nile, Part 2." *Blackwood's Magazine*, October 1859, 391–419.

———. "Captain Speke's Discovery of the Victoria Nyanza Lake, the Supposed Source of the Nile, Part 3." *Blackwood's Magazine*, November 1859, 565–90.

———. "Journal of a Cruise on the Tanganyika Lake." *Blackwood's Magazine*, September 1859, 339–90.

———. *Journal of the Discovery of the Source of the Nile*. Edinburgh: Blackwood & Sons, 1863.

———. *What Led to the Discovery of the Source of the Nile*. Edinburgh: Blackwood & Sons, 1864.

Spiker, Christina M. "'Civilized' Men and 'Superstitious' Women: Visualizing the Hokkaido Ainu in Isabella Bird's *Unbeaten Tracks*, 1880." In *Gender, Continuity, and the Shaping of Modernity in the Arts of East Asia, 16th–20th Centuries*, edited by Kirsten L. Chiem and Laura Blanchard, 287–316. Leiden: Brill, 2017.

Spivak, Gayatri Chakravorty. "Can the Subaltern Speak?" In *Marxism and the Interpretation of Culture*, edited by Cary Nelson and Lawrence Grossberg, 271–316. Basingstoke: Macmillan, 1988.

Spufford, Francis. *I May Be Some Time: Ice and the English Imagination*. London: Faber & Faber, 1996.

Stanley, Henry M. *How I Found Livingstone: Travels, Adventures, and Discoveries in Central Africa*. London: Sampson, Low & Marson, 1872.

———. *In Darkest Africa: Quest, Rescue and, Retreat of Emin Governor of Equatoria*. 2 vols. London: Sampson, Low & Marson, 1890.

———. *My Kalulu, Prince, King, and Slave: A Story of Central Africa*. London: Sampson, Low, Marston & Searle, 1873.

———. *Through the Dark Continent; The Source of the Nile around the Great Lakes of Equatorial Africa and down the Livingstone River to the Atlantic Ocean*. 2 vols. London: Sampson Low, Marston, Searle & Rivington, 1878.

Stanley, Marni. "Skirting the Issues: Addressing and Dressing in Victorian Women's Travel Narratives." *Victorian Review* 23, no. 2 (1997): 147–67.

Stern, Philip. "Exploration and Enlightenment." In *Reinterpreting Exploration: The West in the World*, edited by Dane Kennedy, 54–79. Oxford: Oxford University Press, 2014.

Stings, Sabrina. *Fearing the Black Body: The Racial Origins of Fat Phobia*. New York: New York University Press, 2019.

Stisted, Georgiana M. "Reminiscences of Sir Richard Burton." *Temple Bar*, July 1891, 335–42.

———. *The True Life of Captain Sir Richard F. Burton*. London: H. S. Nichols, 1896.

Stocking, George. *Race, Culture, and Evolution: Essays in the History of Anthropology*. Chicago: University of Chicago Press, 1982.

———. *Victorian Anthropology*. New York: Free Press, 1987.

Stoddart, Anna. *The Life of Isabella Bird: (Mrs Bishop)*. London: John Murray, 1906.

Stoddart, David R. "Do We Need a Feminist Historiography of Geography—And If We Do, What Should It Be?" *Transactions of the Institute of British Geographers* 16, no. 3 (1991): 484–87.

———. "Geography, Education, and Research." *Geographical Journal* 147, no. 3 (1981): 287–97.

Stoler, Ann Laura. *Along the Archival Grain: Epistemic Anxieties and Colonial Common Sense*. Princeton NJ: Princeton University Press, 2009.

———. *Carnal Knowledge and Imperial Power: Race and the Intimate in Colonial Rule*. Berkeley: University of California Press, 2010.

———. *Duress: Imperial Durabilities in Our Times*. Durham NC: Duke University Press, 2016.

———. *Race, Education, and Desire: Foucault's* History of Sexuality *and the Colonial Order of Things*. Durham NC: Duke University Press, 1995.

Strange, Carolyn. "Reconsidering the 'Tragic' Scott Expedition: Cheerful Masculine Home-Making in Antarctica, 1910–1913." *Journal of Social History* 46, no. 1 (2012): 66–88.

Stuart, John McDouall. *Explorations in Australia: The Journals of John McDouall Stuart during the Years 1858, 1859, 1860, 1861 & 1862, When He Fixed the Centre of the Continent and Successfully Crossed It from Sea to Sea*. London: Saunders, Otley, 1865.

Sullivan, Rob. *The Geography of the Everyday: Toward an Understanding of the Given*. Athens: University of Georgia Press, 2017.

Sutherland, John. *Orwell's Nose: A Pathological Biography*. London: Reaktion, 2016.

Swart, Sandra. *Riding High: Horses, Humans, and History in South Africa*. New York: New York University Press, 2010.

Sylph, Ann. "Women in Zoology: Moring into the 20th Century." *Journal of Zoology* (2019). https://zslpublications.onlinelibrary.wiley.com/doi/toc/10.1111/(issn)1469-7998.women_in_zoology (accessed May 3, 2024).

Thiess, Derek J. "Dan Simmons's *The Terror*, Inuit 'Legend,' and the Embodied Horrors of History." *Journal of the Fantastic in the Arts* 29, no. 2 (102) (2018): 222–41.

Thomas, Martin. "What Is an Expedition? An Introduction." In *Expedition into Empire: Exploratory Journeys and the Making of the Modern World*, edited by Martin Thomas, 1–24. London: Routledge, 2015.

Thompsell, Angela. *Hunting Africa: British Sport, African Knowledge, and the Nature of Empire*. Basingstoke: Palgrave Macmillan, 2015.

Thompson, Carl. *The Suffering Traveller and the Romantic Imagination*. Oxford: Oxford University Press, 2007.

Thomson, Joseph. *To the Central African Lakes and Back: A Narrative of the Royal Geographical Society's East Central African Expedition*. 2 vols. London: Sampson, Low, Marston, Searle & Rivington, 1881.

Topdar, Sudipa. "The Corporeal Empire: Physical Education and Politicising Children's Bodies in Late Colonial Bengal." *Gender & History* 29, no. 1 (2017): 176–97.

Tuck, Eve, and K. Wayne Yang. "Decolonization Is Not a Metaphor." *Decolonization: Indigeneity, Education & Society* 1, no. 1 (2012): 1–40.

Tucker, Faith and John Horton. "'The Show Must Go On!' Fieldwork, Mental Health, and Wellbeing in Geography, Earth, and Environmental Sciences." *Area* 51, no. 1 (2019): 84–93.

Tuhiwai Smith, Linda. *Decolonising Methodologies: Research and Indigenous Peoples*. London: Zed, 1999.

Turner, Bryan S. *The Body and Society: Explorations in Social Theory*. 3rd ed. London: SAGE, 2008.

Twigg, Julia. "Carework as a Form of Bodywork." *Ageing and Society* 20, no. 4 (July 2000): 389–411.

Tylor, Edward Burnett. "Anthropology." In *Hints to Travellers: Scientific and General*, 5th ed, edited by Henry Godwin Austen, John Knox Laughton, and Douglas Freshfield, 222–43. London: Royal Geographical Society, 1883.

Uchiyama, Hidehiko, Nobuyo Ohtani, and Mitsauki Ohta. "Three-Dimensional Analysis of Horse and Human Gaits in Therapeutic Riding." *Applied Animal Behavioral Science* 135, no. 31 (2011): 271–76.

Urbanik, Julie. *Placing Animals: An Introduction to the Geography of Human-Animal Relations*. Lanham MD: Rowman & Littlefield, 2012.

Wadiwel, Dinesh Joseph. "Thick Hides: Whipping, Biopolitics, and the White Soul of Power." *Social Semiotics* 19, no. 1 (2009): 47–57.

Warrior, Claire, and John McAleer. "Objects of Exploration: Expanding the Horizons of Maritime History." In *Geography Technology and Instruments of Exploration*, edited by Fraser MacDonald and Charles W. J. Withers, 97–118. Farnham: Ashgate 2015.

Weems, John E. *Peary, the Explorer and the Man: Based on His Personal Papers.* Los Angeles: J. P. Tarcher, 1988.

Wess, Jane A., and Charles W. J. Withers. "Instrument Provision and Geographical Science: The Work of the Royal Geographical Society, 1830–ca. 1930." *Notes and Records: The Royal Society Journal of the History of Science* 73, no. 2 (2019): 223–41.

Whatmore, Sarah. *Hybrid Geographies: Natures, Cultures, Spaces.* London: SAGE, 2002.

Wheeler, Sara. *Cherry: A Life of Apsley Cherry Garrard.* London: Jonathan Cape, 2001.

———. *Terra Incognita: Travels in Antarctica.* London: Vintage, 1997.

Wilks, Sarah L. "The Coldest Dog and Pony Show on Earth: Animal Welfare on the First Expeditions to Reach the South Pole." *Anthrozoös* 25, no. 1 (2012): 93–109.

Williams, Isobel. *Captain Scott's Invaluable Assistant: Edgar Evans.* Cheltenham: History Press, 2012.

Williams, Joe, and Whitney Love. "Low-Carbon Research and Teaching in Geography: Pathways and Perspectives." *Professional Geographer* 74, no. 1 (January 2, 2022): 41–51.

Williams, Laurence. "'Like the Ladies of Europe?' Female Emancipation and the 'Scale of Civilisation' in Women's Writing on Japan, 1840–1880." *Studies in Travel Writing* 21, no. 1 (2017): 17–32.

Wilson, Edward A. *Diary of the* Discovery *Expedition to the Antarctic Regions 1901 1904.* Edited from the original manuscript by Ann Savours. London: Blandford, 1966.

———. "The Medical Aspect of the *Discovery*'s Voyage to the Antarctic." *British Medical Journal* 2, no. 2323 (July 8, 1905): 77–80.

Wisnicki, Adrian S. "Cartographical Quandaries: The Limits of Knowledge Production in Burton and Speke's Search for the Source of the River Nile." *History in Africa* 35 (2008): 455–79.

———. "Charting the Frontier: Indigenous Geography, Arab-Nyamwezi Caravans, and the East African Expedition of 1856–59." *Victorian Studies* 51, no. 1 (2008): 103–37.

———. *Fieldwork of Empire, 1840–1900: Intercultural Dynamics in the Production of British Expeditionary Literature.* London: Routledge, 2019.

Withers, Charles W. J. "Disguise—Trust and Truth in Travel Writing." *Terrae Incognitae* 53, no. 1 (2021): 48–64.

———. "Geography and 'Thing Knowledge': Instrument Epistemology, Failure, and Narratives of Nineteenth-Century Exploration." *Transactions of the Institute of British Geographers* 44, no. 4 (2019): 676–91.

———. *Majestic River: Mungo Park and the Exploration of the Niger*. Edinburgh: Birlinn, 2022.

———. "Mapping the Niger, 1798–1832: Trust, Testimony, and 'Ocular Demonstration' in the Late Enlightenment." *Imago Mundi* 56, no. 2 (July 2004): 170–93.

———. "Memory and the History of Geographical Knowledge: The Commemoration of Mungo Park, African Explorer." *Journal of Historical Geography* 30, no. 2 (2004): 316–39.

———. *Placing the Enlightenment: Thinking Geographical about the Age of Reason*. Chicago: University of Chicago Press, 2007.

———. "Trust—in Geography." *Progress in Human Geography* 42, no. 4 (2018): 489–508.

Worboys, Michael, Julie-Marie Strange, and Neil Pemberton. *The Invention of the Modern Dog: Breed and Blood in Victorian Britain*. Baltimore MD: John Hopkins University Press 2018.

Wulf, Andrea. *The Invention of Nature: The Adventures of Alexander von Humboldt, the Lost Hero of Science*. London: John Murray, 2015.

Wylie, John. "Becoming-icy: Scott and Amundsen's Polar Voyages: 1910–1913." *cultural geographies* 9, no. 3 (2002): 249–65.

———. *Landscape*. Abingdon: Routledge, 2007.

Wyndham-Quin, Windham Thomas. *Past Times and Pastimes*. 2 vols. London: Hodder & Stoughton, 1922.

Young, Donald, ed. *The Search for the Source of the Nile*. London: Roxburghe Club, 1999.

Yusoff, Kathryn. "Antarctic Exposure: Archives of the Feeling Body." *cultural geographies* 14, no. 2 (2007): 211–33.

———. "Climates of Sight: Mistaken Visibilities, Mirages, and 'Seeing Beyond' in Antarctica." In *High Places: Cultural Geographies of Mountains, Ice, and Science*, edited by Denis Cosgrove and Veronica della Dora, 48–63. London: I. B. Tauris, 2009.

———. "Configuring the Field: Photography in Early Twentieth-Century Antarctic Exploration." In *New Spaces of Exploration: Geographies of Discovery in the Twentieth Century*, edited by Simon Naylor and James R. Ryan, 52–77. London: I. B. Tauris, 2010.

Zambas, Chantal. *Out in Africa: Same-Sex Desire in Sub-Saharan Literatures and Culture*. Oxford: James Currey, 2013.
Zweiniger-Bargielowska, Ina. *Managing the Body: Beauty, Health, and Fitness in Britain, 1880–1939*. Oxford: Oxford University Press, 2010.

INDEX

Page numbers in italics indicate illustrations.

www.ingramcontent.com/pod-product-compliance
Lightning Source LLC
Chambersburg PA
CBHW021712090525
26443CB00002B/74
9781496230973